MEANINGFUL INEFFICIENCIES

MEANINGFUL INEFFICIENCIES

Civic Design in an Age of Digital Expediency

Eric Gordon

and

Gabriel Mugar

OXFORD
UNIVERSITY PRESS

OXFORD
UNIVERSITY PRESS

Oxford University Press is a department of the University of Oxford. It furthers
the University's objective of excellence in research, scholarship, and education
by publishing worldwide. Oxford is a registered trade mark of Oxford University
Press in the UK and certain other countries.

Published in the United States of America by Oxford University Press
198 Madison Avenue, New York, NY 10016, United States of America.

Library of Congress Cataloging-in-Publication Data
Names: Gordon, Eric, 1973– author. | Mugar, Gabriel, author.
Title: Meaningful inefficiencies: civic design in an age of
digital expediency / Eric Gordon and Gabriel Mugar.
Description: New York, NY : Oxford University Press, [2020] |
Includes bibliographical references and index.
Identifiers: LCCN 2019035302 (print) | LCCN 2019035303 (ebook) |
ISBN 9780190870133 (paperback) | ISBN 9780190870140 (hardback) |
ISBN 9780190870164 (epub) | ISBN 9780190870157 (updf) |
ISBN 9780190870171 (online)
Subjects: LCSH: Urban policy—United States—Citizen participation. |
Human services—United States—Planning—Citizen participation. |
Municipal government—United States—Citizen participation. |
Community development—United States. | Trust—United States.
Classification: LCC HT123 .G579 2020 (print) | LCC HT123 (ebook) |
DDC 307.760973—dc23
LC record available at https://lccn.loc.gov/2019035302
LC ebook record available at https://lccn.loc.gov/2019035303

1 3 5 7 9 8 6 4 2

Paperback printed by Marquis, Canada
Hardback printer by Bridgeport National Bindery, Inc., United States of America

Dedicated to the memory and spirit of Moses Shumow

(1977–2019)

CONTENTS

ACKNOWLEDGMENTS

This book is the outcome of years of research, conversations both formal and informal, and feedback from so many with a shared interest in the work. There are probably too many to list here, but we're going to try anyway. Thank you to everyone who participated in the interviews that make up the cases in this book. Your honesty, generosity, and commitment to your work inspired us, and we hope this book does justice to your vision as civic designers. Thank you to those who helped with the research or provided feedback on early drafts of the report that became this book: Jessica Weaver, Jesse Fryburg, Mariam Chahine, Erhardt Graeff, John Gastil, Jennifer Bradley, Laura Forlano, Joe Kahne, Ceasar McDowell, and Sean Van Deuren. Thank you to Jennifer Manuel and Daniel Lambton-Howard from Newcastle University for their work on developing the interview protocol. Andrew Divigal and Regina Lawrence at the Agora Journalism Center at the University of Oregon were supporters and collaborators in our research with journalists. And we want to acknowledge the MacArthur Foundation and the Knight Foundation for supporting various parts of the research. Then there were those who, out of the goodness of their hearts, read early drafts of the book's chapters. Thank you to Benjamin Stokes, Carl DiSalvo, Peter Levine, jessika maria ross, Justeen Hyde, Regina Lawrence, and Miguel Sicart. Your feedback was invaluable as we revised the manuscript.

We couldn't have written this book without the brilliance and inspiration of Stephen Walter, who contributed so much to the spirit of this book through long conversations, collaborations, and gracious sharing of his depth of knowledge. And he was there, along with Jesse Baldwin-Philippi, when we first uttered the phrase *meaningful inefficiencies* and immediately scribbled it on a whiteboard. We knew it was interesting long before we actually knew what it meant. And thank you to Nigel Jacob, a close friend and fellow traveler, who taught us what government innovation could look like and gave us opportunities to work with a variety of stakeholders in Boston and beyond.

There are others who provided support, either directly or indirectly, to the formation of the ideas in this manuscript. Ethan Zuckerman provided us with a persistent critical voice, if only as an imagined angel (or devil?) on our shoulders pushing us to think differently. Catherine D'Ignazio and Paul Mihailidis, two of the most inspiring colleagues one could ask for, listened to us wax poetic about civic design and weren't afraid to point out what we were missing. Paula Ellis is a friend and inspiration whose practical frame for complex ideas inspired us to always think about application. Robert Sabal, dean of Emerson College's School of the Arts, provided moral and structural support. And then there are our graduate students in the Civic Media Art and Practice program (now called Media Design) at Emerson College. They engaged in these ideas when they were mere scattered lectures and provided thoughtful feedback that always pushed us to do better. Special thanks to Eric's graduate assistants Tijana Zderic who helped create the index and Kara Jackson who double and triple checked our references.

This book would not have been possible if not for the staff (past and present) at the Engagement Lab, who made it possible for us to be practicing civic designers even as we studied the practice.

Eric wants to thank his wife, Justeen Hyde, and his children, Elliot and Adeline Gordon, who continually push him to be a better person and inspire him to do the work he does.

Gabriel wants to thank his wife, Anda French, with whom he has shared countless explorations around community building, design, and public space. He also wants to thank his son, Oliver, who inspires him to see the world with fresh eyes every day.

INTRODUCTION

In December 2017, nearly 200 parents of Boston Public School children gathered at a school committee meeting to protest the implementation of new start times for schools across the city (see Figure I.1). Many of the city's high schools had extremely early start times—well before 8 AM—and the district was seeking to respond to research that teenagers generally do better later in the morning.[1] The goal was to have all high school students start after 8 AM and all primary school students home before 4 PM. But because of a complex juggling act that required stretching a limited number of school busses across the city's 125 schools, changing some start times meant making changes to nearly every school's start time. The result of some mathematical modeling was new start and end times for 84% of the city's schools, in many cases with over 2-hour differences. Some elementary schools in the city had their start times changed from 9:30 to 7:15 AM. When these new schedules were announced, many parents were furious because they felt they were never consulted. After all, the district's decision would have major consequences on families' sometimes precarious juggling act of drop-off, pickup, and after-school care. The day after the protest, superintendent Tommy Chang announced a reversal of the

Meaningful Inefficiencies. Eric Gordon and Gabriel Mugar, Oxford University Press (2020).
© Oxford University Press.
DOI: 10.1093/oso/9780190870140.001.0001

Figure I.1. Boston Public School parent expresses opposition during a Boston School Committee meeting. (Courtesy of Craig F. Walker / Globe Staff)

policy, saying that they would spend another year considering the implications of the changes.

The school times were calculated by a team of researchers at the Massachusetts Institute of Technology who developed a system to change school schedules in coordination with new bus routes that were put in place the year before. It was a feat of systems engineering that allowed them to meet specific requirements about start and end times while coordinating with the fleet of busses crossing Boston's congested narrow streets. The only thing they didn't appropriately consider was the impact these positive changes for high school students would have on the city's primary school students and families. As angry parents held signs such as "Students Are Not Widgets" and "Families Over Algorithms," the message to the school committee was unmistakable. Big data has its limits, and sometimes the most important data comes from talking to people.

Boston's misstep is just one of many examples in cities across the country, where leaders intoxicated by the promise of data-driven decision making are not able to see the trees through the forest. In theory, this was a triumph of efficiency, defined as the "least expenditure of a limited resource necessary to achieve a given goal" (Suits, 2005, p. 62). It connected two complex systems with the least amount of friction. Until, of course, people spoke up and added friction. In the case of the Boston Public Schools, families were ill served by the efficient solution, largely because they were poorly consulted on defining the problem. While efficiency is often a priority of government organizations, it is not always for the people those organizations serve.

People tend to like efficient systems if the goals are clear, universal, and transactional—for example, a new city website to make it easier to pay parking tickets or a new distribution of salt trucks before a snowstorm to cover more streets faster. These efficiencies would likely be welcomed without need for extensive consultation. But when decisions about systems impact the structure and flow of people's everyday lives, the justification for boundaries and rules matters, and friction becomes a feature—not a bug.

A common assumption made by designers of all sorts is that users want a seamless interaction with the process, tool, or program they are designing. According to the computer scientist Mark Weiser (1993):

A good tool is an invisible tool. By invisible, I mean that the tool does not intrude on your consciousness; you focus on the task, not the tool. Eyeglasses are a good tool—you look at the world, not the eyeglasses. The blind man tapping the cane feels the street, not the cane. Of course, tools are not invisible in

themselves, but as part of a context of use. With enough practice
we can make many apparently difficult things disappear. (p. xx)

Weiser inspired a generation of computer scientists and technologists
to see computing in context, to approach the design of technological systems as deeply embedded within social and physical systems.
Like eyeglasses, good technology blends into the background—
becomes entirely frictionless—so as to highlight the social and
physical realities over the technology that enables it. The philosopher Martin Heidegger, in one of his most famous neologisms,
introduced the concept of the "ready-to-hand," an attitude toward
technology wherein it is treated as an extension of the human body,
not an object of contemplation. Ubiquitous computing is an aspi-
rational framework for technologists to imagine and design for this
seamless extension (Dourish, 2014). This vision of technology has
informed, inspired, and reflected decades of design, from the iPhone
to Uber, and from city websites to data dashboards. The promise
of frictionless technological systems has been a driving force in the
space of innovation, from the tech sector to the social sector.

But too often, as in the case of the Boston Public Schools, that
frictionless design excludes necessary consultation, ultimately
causing the system to break. This is most apparent when the values
underlying the technological system are in conflict with the values
underlying the social system. When parents resisted the school
committee's decision, seemingly dictated by machines, they were
not resisting the outcomes *per se,* but the mechanics of decision
making. The optics of machine intelligence determining what is
best for children and families were anathema to commonly held
assumptions or aspirations of democratic processes. In this case,
the invisibility of computing was shrouding the desired visibility of
democracy.

This book is a reflection on a contemporary moment, where networked technology—specifically big data and its seamless interactions—is having a significant impact on the shape of public space and public decision making. This is not a book about the dangers of technology. We do not advocate for a media blackout or a retreat to analog processes. Instead, we advocate for a consideration of the values that technology possesses and how these values commingle with the systems they are augmenting. When groups strive for "smart cities" or "smart government," they are bringing normative assumptions about the values of smart technologies to bear on systems such as urbanism and democracy (de Lange & de Waal, 2013). This creates a tension between visibility and invisibility, between friction and frictionlessness, and between efficiency and inefficiency.

We examine how government, civil society, and journalism organizations are struggling with these contradictions. When smart tracking and clever algorithms are on offer to enhance service delivery and increase efficiency, how do organizations think about human interaction? Regardless of how technologically advanced an organization is, each is negotiating a shift in perception and expectation about how things should get done. We draw on examples from a range of organizations that we characterize as public-serving organizations so as to speak across domains and disciplines. There are more similarities than differences in how practitioners across these spaces are responding to a contemporary moment. For most public-serving organizations, technological progress is seen as necessary for continued relevance (Gordon & Lopez, 2019). But the practice of adapting technologies for local problem solving can be challenging. It's never as simple as adopting the shiny new tool. It requires a rethinking of infrastructure (Bowker et al., 2010; Rhinesmith, 2016), a consideration of how that tool interfaces with existing practices and social norms. Adopting new public-facing technologies is typically as much about internal shifts and the rethinking of process as it is about external communication. New, connected technologies are a double-edged sword: on one edge, they hold the promise of efficient solutions, and on the other, the threat of obfuscating or bypassing the process and relations that generate solutions.

What's at stake in this contemporary transformation is the ability of organizations to mediate trusting relationships with publics. A recent study from the Pew Research Center has shown that trust in the US federal government is at an all-time low (Pew Research Center, 2017). The marketing firm Edelman releases a "trust index" every year. Their 2018 report shows trust in a range of institutions globally (from local government to media) stabilizing after a rapid downturn in 2017. The main exception is in the United States, where trust dropped 23%—the biggest drop in the 17 years they have conducted the survey. Individuals do not trust in institutions as much as they once did. They have more information and less understanding of how to authenticate it (Zuckerman 2017). With Russian hacking into political process in the United States and elsewhere and a rise in strongman politicians around the globe regularly questioning the legitimacy of the press when it disagrees with them, there is good reason for active citizens to question the intentions of the faceless institutions that mediate public life. The context is complicated: Public trust in the institutions mediating civic life is low, there is an assumption that greater efficiency will build trust, and organizations are undergoing significant infrastructural and programmatic changes through adopting new technologies (Morison, 2010).

In some cases, when an underperforming organization adopts technology to enhance its output, people begin to trust in that organization's ability to do its job (Harding et al., 2015). When a city updates its website to enhance usability, or when online payments are streamlined, better user experience typically results in higher trust (Porembescu, 2016). But when a city installs kiosks that capture IP addresses of passersby without any input from residents, or when black box algorithms determine what news content you see on your browser, the absence of process can have the opposite impact. Efficiency, in the sense of charting a path to a goal with the least amount of friction, can be at odds with the goal of building trust in the institutions that mediate public life. In general, public-serving organizations seek a balance between transactional and relational models of getting things done. However, the promise of new technologies and the rush to implementation is creating a lopsidedness. As new digital tools compel organizations toward the transactional, and as publics grow increasingly distrustful of the role of civic institutions broadly, there is now more need than ever to achieve balance.

This book is about those practices that challenge the normative applications of "smart technologies" in order to build or repair trust with publics. It is a book about design, but not necessarily about designers. It is about those people working within public-serving organizations that are attempting to reshape programs, mission, and purpose by creating the conditions through which organizations form relationships and build trust with publics. Without coordinating, these designers have adopted a common framework we call "meaningful inefficiencies," or the deliberate design of less efficient over more efficient means of achieving some ends. These practitioners are already employed by organizations all over the world; but in most cases, the organizations don't know it yet. They masquerade as technologists, communication specialists, journalists, producers, and engagement officers, but they're doing the work of designers, and they are thoughtfully, and often quietly, innovating the shape of civic life.

This book draws attention to these practices and encourages a rethinking of how innovation within public-serving organizations is understood, applied, and sought after. As more and more technologies are applied to the ins and outs of civic life, from voting to consultation with elected officials, to advocating for local change, we remain enthusiastic about the possibilities of enhanced efficiency. But we are confident that unchecked efficiency can be damaging for public-serving organizations to achieve their desired goals, specifically when they require trust and care from the publics they serve.

DESIGN

Design scholar Chris LeDantec (2016) uses the term "social design" to describe the process of designing technologies with people who become invested in the outcomes. What we call civic design is social, as LeDantec describes it, but it also refers to the intended outcome of the design process—the creation of civic spaces. The practitioners we call civic designers are creating opportunities to interact, form alliances, generate shared interest, and care for matters of public concern. There are a growing number of these practitioners in the United States and beyond. Political philosopher Peter Levine in his book *We Are the Ones We Have Been Waiting For* (2015) documents this growth in the United States. He calls for a daylighting of the 1 million most active citizens who are pursuing local, deliberative, and novel approaches to civic life into a self-conscious civic renewal movement. Levine's book is a smart and optimistic articulation of a contemporary moment where there is an emergent commitment to recognizing and supporting a plurality of publics through deliberation and discourse. And while Levine focuses on the broad strokes of the political moment, we turn our attention to the texture of the work being done and the challenges of the individual practitioners.

The civic designers on whom we focus in this book are creating meaningfully inefficient spaces. This practice entails creating systems with clear rules and outcomes wherein the participant is able to learn, connect, and engage with a system, all for the purpose of building trust and engagement. So what makes a process *meaningfully* inefficient? It's when the inefficiency scaffolds interactions that serve to connect people to one another. Consider the example of a remote-controlled car. For many children, it is not an uncommon experience to spend hours putting together a car, only to play with it for 30 minutes before putting it in the closet for the remainder of its existence. Building a model remote-controlled car can be an enjoyable process for children and adults, but one typically rushes through building it to arrive quickly at the goal. As a counterpoint to this phenomenon, the Nitro Racer is a model remote-controlled car that comes as part of a magazine subscription, where each month exactly one part is delivered with the issue. After 3 years, the final part arrives and the car can be completed. This would be a mere inefficiency if the goal was simply to play with the car. But if the goal, as it was explained to us by a colleague, was to have an extended project to do with his son, then in fact the inefficiency was meaningful. The self-imposed obstacle of the 3-year timeline transforms the goal of the project from the completion of the car to the time spent with someone while completing the car (see Figure I.2). The Nitro Racer is not an innovation in model cars; it is an innovation of the system one engages in to build a model car. This is an important distinction, because innovative design does not always mean innovative objects; it can mean innovative logics through which objects are produced.

Figure I.2. Cover of Nitro Racer magazine.

When a system is designed as a meaningful inefficiency, it is placed in contrast to normative assumptions of the value of efficiency. We first arrived at this insight through our own experience designing games for public-sector organizations. Both of us were affiliated with the Engagement Lab, an applied research lab at Emerson College in Boston, known for its work in creating civic games. As our work became recognized, more and more organizations approached us to "design a game for *x* or to make people do *y* with a game." Those wanting games for their organization saw them largely as motivation systems to encourage behaviors or reinforce knowledge; with all the attention paid to outcomes, rarely did they notice the benefits that emerged simply from playing the game. Games rarely make things move faster—they function like the Nitro Racer, placing unnecessary obstacles in one's way so that players might find meaning in the process.

The decision a player makes to step outside of everyday life and to step into a play space, or magic circle (Huizinga, 1955), is almost always an inefficiency inserted into some larger system of exchange and interaction. The philosopher Bernard Suits describes the experience of playing a game as follows: "To play a game is to attempt to achieve a specific state of affairs [prelusory goal], using only means permitted by rules [lusory means], where the rules prohibit use of more efficient in favour of less efficient means [constitutive rules], and where the rules are accepted just because they make possible such activity [lusory attitude]" (2005, p. 10). A prelusory goal is typically identified by asking the question "How do I win?" This is a goal defined for the player before she enters into the game system. I win in chess by capturing my opponent's king. Once I understand that, I can enter into the "magic circle of the game," which is defined by a set of rules and limitations (prelusory means), where those rules stop me from achieving my goal (capturing my opponent's king) in the most efficient means possible (reaching across the board and grabbing it). These constitutive rules, as Suits calls them, provide the restrictions and limitations that are necessary for play. And interestingly, players accept these rules precisely because they allow them to play. As such, *games are, by definition, inefficient*; play requires the space that contained inefficiency affords. Put another way, "meaningful inefficiencies" is the productive lag in systems generated by rules that enforces and justifies *playing* (Gordon & Walter, 2016).

This book is inspired by games as they are a productive analog to the kinds of play spaces created by civic designers. And while we will be using actual games as examples throughout the book, games are by no means a privileged or ideal form. They are one among many objects of design that practitioners are using to correct for smart solutions. We are not interested in wrenches in the system, or external objects inserted into efficient systems to momentarily slow them down. In the urban context, these kinds of practices are often called "tactical," or "acupuncture" (Matchar, 2015), borrowing from the tradition of Guy Debord's *Psychogeography* (2000). Consider a public space intervention like Park(ing) Day that, once a year, organizes thousands of people around the world to transform parking spaces into mini, temporary parks (see Figure I.3).

Park(ing) Day encourages people to stop and use space differently. This extraordinary event, which started in San Francisco in 2005 as a public art installation by the art collective Rebar, has since become something of a global holiday. Rebar describes Park(ing) Day as a prototype in open source urban design and "a global experiment in remixing, reclaiming and reprogramming vehicular space

Figure I.3. Park(ing) Day in Cambridge, Massachusetts.

for social exchange, recreation and artistic expression" (Rebar 2011, p. 1). While we are interested in this kind of work, our focus is on design interventions that seek to transform systems or institutions by building sustainable inefficiencies within them. Park(ing) Day is a novelty. It happens annually and then the system of parking and vehicular and pedestrian mobility returns to the status quo. A meaningful inefficiency, on the other hand, is a way of reconfiguring systems. To design meaningful inefficiencies is to design spaces for play within a larger system, such that the consequences of play augment the player's perception and use of the system. For example, participatory budgeting can be a meaningful inefficiency, insofar as it presents a goal and inserts lengthy deliberation as an obstacle to achieving that goal (Lerner, 2015). The result is not merely to slow down the budgeting process, but to open up a piece of the process to public deliberation so as to both instruct the public in how budgeting works and to empower them in shaping outcomes. The outcome of meaningful inefficiencies in these examples is to hold space for people, for social interaction, and for moments of productive encounter in which infinite possibilities exist. Without these encounters, organizations dictate a world prescribed by systems, predicted by their makers with little room for variability.

Most public-serving organizations are eager to adopt new technologies or tools that "cut the fat" or "reduce friction" in the system, not the other way around. In our research with practitioners, we acknowledge this tension and recognize a difference between a meaningful inefficiency and a *mere* inefficiency. When a website loads slowly, or one has to wait in three different lines at the Department of Motor Vehicles to renew a driver's license, these *mere* inefficiencies mostly lead to anxiety and frustration for users. In the rush to address these inefficiencies, especially when introduced to digital tools that are seemingly designed for such purpose, organizations can easily design against the values of the publics they seek to cultivate. In the example of the Boston Public Schools, the clever engineering solution optimized bus scheduling but exacerbated the public's mistrust. It solved one problem and made another worse. Incorporating dialogue, opportunities for the public to explore the algorithm, and investing in building the public's capacity to understand the data being used are all meaningful inefficiencies appropriate to the design for and with publics.

FRAMING

This book is based on years of practice designing with public-serving organizations and over 60 interviews, and self-administered reflections, conducted by and with practitioners invested in building strong publics through media and technology. All interviews and reflections were conducted between 2013 and 2018. All interviews were conducted under the supervision of an Institutional Review Board (IRB), but they were collected as part of four distinct research projects. For that reason, some names and identities are withheld and some are not. This is the outcome of how the data was collected and has nothing to do with the nature of the data itself. In addition to these interviews and reflections, we draw from our direct experience doing participatory design and research in contexts around the world. Each of the authors brings unique organizational perspectives: Eric Gordon is a professor and director of the Engagement Lab, and Gabriel Mugar is a design researcher with IDEO, a major for-profit design consultancy. Although much of the research was conducted while we were both at the Engagement Lab, as we write this book together, we approach design from different organizational contexts (academia and private sector), hopefully bringing a rich nuance to the conclusions and practical recommendations of the book.

The majority of our inputs focus on how individual actors are navigating institutional structures and circumventing obstacles as they work with media and technology. As a result, this book places a practical emphasis on organizational and institutional transformation. We are inspired by the work of philosopher Hugh Heclo. In his book *On Thinking Institutionally* (2008), he argues that institutions are not legal structures; they are norms that guide how groups of individuals act collectively. He suggests that institutions frame most social interactions whether or not people are aware of them. He goes on to criticize what he deems the "postmodern stance," which rejects all inherited values as cultural oppressions and believes that "meaning is to be found only in self-creation, not faithful reception of something beyond oneself" (pp. 100–101). Heclo argues this position mistakes the organization for the institutional values that underlie it. Every individual or individual organization brings the moral framework of institutions to bear on their actions. A nonprofit, local news organization is operating in the institutional framework of journalism, even if they resist the dominant values that comprise it.

So as not to fall into that "postmodern stance," wherein a false binary is created between self and institution, Heclo introduces the concept of "thinking institutionally," which "is to enter and participate in a world of larger, self-transcendent meanings" (p. 107). Thinking institutionally is not a matter of supporting or rejecting existing institutional logics, but in thinking through (either critically or otherwise) the values or moral obligations that undergird institutions and the way these morals and values are performed.

While it is commonly understood that organizations, especially large ones, can be slow to change, they are always comprised of actors that are negotiating institutional values with organizational hurdles or obstructions. These individual actors borrow from some larger institutional framework (be it democratic governance or news) and are guided by moral obligations that correspond to democratic values, such as inclusion, equality, and collective responsibility. Practice, or the things that people actually do in their lives, is a constant negotiation between what needs to get done and the values informing the institution.

What we witnessed in organizations is that engaging in the work of meaningful inefficiencies requires thinking institutionally. The work of facilitating democratic process through trust building is more than just an individual action, or even the action of a single organization. It requires forefronting institutional values, even if they are not clearly stated and universally accepted. This expression of values is often in conflict with the efficiency needs of an organization. For example, a grassroots news organization in Chicago invests time in facilitating community conversations around controversial police data, as opposed to a singular focus on publishing more stories. The reach is reduced because of the investment in community building, but the practice represents an investment in relationships and a mind toward sustaining this work for the long term. In previous research, we have called this values-forward media work "civic media." In the introduction to the edited collection *Civic Media: Technology, Design, Practice*, Gordon and Mihailidis define the term as "any mediated practice that enables a community to imagine themselves as being connected, not through achieving, but through striving for common good" (2016, p. 2). There are two important aspects of this definition: (1) "striving for" suggests process over product, and (2) "common good" suggests a shared set of negotiated values driving the work. Before every finished product, before every celebrated new initiative, values, interests, and power dynamics must be navigated and negotiated. This negotiation is foundational to any group of people called a public. We reject the notion of a singular public (the public), and we explore in great detail how publics form in the creation of meaningful inefficiencies and as the subject of the designed system (see Chapter 2).

Civic design, as we define it, requires a challenging of the presumptions of efficiency; namely, a challenge to the presumption that the primary purpose of designed systems is to achieve a stated goal with the least expenditure of resources. The insertion of slack, or room to play, within a system is a necessary action to achieve this (a concept that we will explore in great detail in Chapter 3). The reconsideration of the object or goal, however, is equally as important. The object of the civic design process is not a new program or a new tool. It is *care* (see chapter 4).

There is a rich literature in philosophy and political science that explores the notion of care. Martin Heidegger speaks of care as a fundamental way of being in the world, from attending to or making use of something to investigating something, care is synonymous with presence. For political philosophers Fisher and Tronto, caring is much more action oriented. It is a "species of activity that includes everything that we do to maintain, continue, and repair our 'world' so that we can live in it as well as possible" (1990, p. 40). For Tronto (1993), care is more than a private moral value; it is an essential part of citizenship in a democracy, orienting people toward an understanding that citizenship is the practice of how one works with others to take care of the world they live in. Within this context, Tronto asks: "How can people claim to live in a democracy if their fears and insecurities begin to override their abilities to act for the common good?" (2013, p. 6). She associates acting for a common good with the act of caring for others, and she argues that democracy is about assigning caring responsibilities.

Tronto defines a hierarchy of caring responsibilities, from caring about, which suggests an attentiveness to a person or issue (this was Heidegger's concern); caring for, which implies a relation and reciprocity; caregiving, which implies the actual action; and care receiving, which is the response to the action. And she proposes a fifth stage that she calls "caring with." "The final phase of care requires that caring needs and the ways in which they are met need to be consistent with democratic commitments to justice, equality and freedom for all" (p. 23). She explains further that this feminist democratic care ethic "is relational." By this view, "the world consists not of individuals who are the starting point for intellectual reflection, but of humans who are always in relation with others" (p. 36). And she argues that the real work of democracy is not caring *per se*, but the assigning of caring responsibilities. In other words, democracy is the negotiation of how giving and receiving care is distributed throughout a society. For our purposes, care is always the outcome of civic design—it is what publics, when given the opportunity to create together, seek to achieve. As we will explain in Chapter 4, caring *with* is not some unachievable democratic ideal, but can be the everyday experience of being in civic systems that foreground democratic values.

Meaningful inefficiencies reinforce what the philosopher Hannah Arendt would call the human condition. The human condition, she says, exists within a tripartite formulation known as the *vita activa*, which includes labor, work, and action. The *vita activa*, the active life, distinct from the *vita contemplativa*, or the thinking life, is what distinguishes humans from animals. Moving away from the Cartesian *cogito ergo sum* (I think therefore I am), the *vita activa* suggests that humans are active, purpose and meaning driven, and defined by labor, not just thoughts. Labor, that which is ongoing and meets biological needs for subsistence, is put into relation with work, or that which has specific outcomes and meaning. And each is in relation to action, which is the means by which humans disclose themselves to each other. Action can be a conversation with friends, a post on social media, or simply co-presence. What's important about Arendt's understanding of action is that it is always open ended—it invites new beginnings—and is connected to labor and work. Our use of the word *action* throughout this book stems from this Arendtian concept. So action is not just a thing one does, but an activity that purposely seeks to unleash new beginnings.

Writing in the 1950s, in direct response to the horrors of World War II, Arendt warns of *dark times,* a world where the opportunities for action taking are drastically reduced. This is a world where hyperrationality justifies behavior, and focus on efficient outcomes distracts from one's sense of a shared world. "Nothing in our time is more dubious than our attitude toward the world," says Arendt (1968, p. 3). The world is all the stuff that lies between people—in particular, the sense of public or society, the sense that one's individuality is contained within a discursive context that provides meaning. She explains: "When men [*sic*] are deprived of the public space—which is constituted by acting together and then fills of its own accord with the events and stories that develop into history—they retreat into their freedom of thought" (1993, p. 5).

Arendt explains that in order to escape, to retreat into our freedom of thought, one needs to reject the world. One turns one's backs on injustice, inequities, sometimes on blatant violence, and looks to the world, not as something with which to contend, but as something one must escape. These individuals choose not to identify with corrupt and corrosive political leaders. "Those who reject such identifications on the part of a hostile world," warns Arendt, "may feel wonderfully superior to the world, but their superiority is then truly no longer of this world; it is the superiority of a more or less *well-equipped cloud-cuckoo land*" (1993, p. 7, emphasis in original).

One retreats into a cloud-cuckoo land, where reality is exactly as one expects it to be—"a shift from the world and its public space to an interior life, or else simply to ignore that world in favor of an imaginary world 'as it ought to be' or as it once upon a time had been" (1993, p. 12). What's more, this imaginary world is *well equipped*, meaning that the structures of society are reinforcing the appropriateness of retreat. Today, where "fake" news is more popular than legitimate news, where confirmation bias compels one to click on what one thinks she already knows, where misinformation perpetuated by the President of the United States sows doubt on science and research, dark times is not simply an historical concern. It is descriptive of a contemporary moment where technology is promising solutions to public problems at such a fast clip that public-serving organizations are having to adopt solutions wholesale without the opportunity to consider consequences. There is no room for publics to form when efficiency is the primary value proposition. Government wants to transform them into customers; news organizations think primarily about clicks. Organizations seek to fade into the background, like Weiser's eyeglasses, and become a conduit to outcomes. But in their attempt to win trust by giving people what they want, they are actually losing trust because the publics they are there to mediate are slowly fading away. Without action, there are only users. Without action, there are no publics. The growing group of practitioners that understand this dynamic are creating systems to transform public-serving organizations.

WHAT PEOPLE ACTUALLY DO

Meaningful inefficiencies is not just public engagement. In government, public engagement is when a decision-making body asks the
general populace for formalized input, outside of a voting booth.
According to Jake Blumgart, this is a relatively new term, first used
by *The New York Times* in 1998 to describe the "political lethargy
surrounding the presidency of Bill Clinton" (2014). Subsequent
to that first usage, it has become rather a popular phenomenon,
leading to a new standard practice in local government, the Obama
administration's Office of Public Engagement, and an entire industry of consultants to support it. While public engagement and
the growing body of professionals working in this space comprise
part of the story, it is not the whole story (Leighninger, 2011). That
public engagement has become compulsory in government decision
making in the United States and Europe is more of a symptom than
a solution to the growing trust deficit. Many of the practitioners we
spoke with would characterize their work as "public engagement,"
but the activities in which they are actually engaged cannot be defined by any existing professional standard or industry practice. The
design of meaningful inefficiencies requires a specific approach, we
argue, that often conflicts with the institutional mandates for more
(not necessarily better) public engagement.

So far, we have described the structure of civic design, but what
of the actual content? Through our interviews with practitioners
and reflections on our own practice, we have identified four primary
activities that comprise a meaningfully inefficient process: network
building, holding space for discussion, distributing ownership, and
persistent input. We first documented these activities in the report
entitled *Civic Media Practice*, where we explored the challenges

practitioners face when actually doing this values-forward work (Gordon & Mugar, 2018). We use these activities throughout this book as a means of describing what people actually do, and in the book's last chapter, we discuss how the activities can be used in the evaluation of such practices, distinct from standard outcome assessment models.

Network building is the act of convening either in person or online for the purpose of social connectivity and solidarity. Such convenings, which can include community centers or social media platforms, support encounters between stakeholders and allow people to identify critical mass around local issues as well as explore possible approaches for taking on particular challenges. These sorts of encounters build networks that further enable opportunities for sharing experiences and knowledge. *Holding space for discussion* is doing the work of assuring that there is time and space for discussion that makes room for multiple viewpoints and is tolerant of dissent. *Distributing ownership* takes place when practitioners outline clear pathways to participation, actively encouraging a power dynamic where stakeholders take the reins of the practice, or when practitioners adopt an open-source ethos to their work, sharing knowledge and encouraging appropriation and repurposing of practice. And *persistent input* is when practitioners not only ask people what they think, but they do so from a position of stability, continuity, and trust: asking once, and then being in the same place to ask again. This persistence is reflected in long-term relationships between practitioners and the communities they work in.

Civic designers spend a huge amount of their time focused on these activities. The activities don't often get documented or reported because there is rarely a clear line drawn between this work and project outcomes and, as a result, no incentives attached to doing it. The work of this book is both to describe these activities so that they become comprehensible to the interested practitioner and to contextualize them within a contemporary moment that is urgently in need of their widespread adoption.

READING THE BOOK

The work of civic design necessitates some freedom of movement, some play. In many, if not most, cases, people within especially large organizations do not have this freedom of movement, and it is immensely difficult to design for it. This is precisely why we foreground those practitioners who have been able to, for any number of reasons or conditions, carve out the time and space to do such work. Their voices are distributed throughout this book. We use some of our own projects as case studies, not because we see ourselves as exemplary actors, but because the freedom afforded to us, especially coming from an academic research unit, allows us to experiment more freely than our colleagues in government or civil society. Additionally, as this book is ultimately about the messiness of practice, and the practitioners we spoke to are largely operating with the best of intentions, we are much more comfortable shining a light on our own work, blemishes and all.

This book is at once deeply philosophical and passionately practical. It is written for the practitioner scholar and the scholarly practitioner. It is a short book, with an argument that builds throughout, so it is meant to be read from beginning to end. Each of the book's chapters is organized conceptually around foundational terms: innovation, publics, play, and care. Essentially, they answer the following questions: What is civic design, for whom, with what, and for what? The final chapter, "Practice," provides practical guidance to implementing civic design within organizations and suggests possible applications beyond public-serving organizations.

In the first chapter on *innovation*, we explain what sets civic innovation apart from other mainstream discourses of innovation. We address concepts of newness, disruption, and novelty, and articulate how they intersect with values of stability and integration. We aim to build constructive tension between the world of technological innovation—Silicon Valley, smart cities, and big tech—and the world of civics, where trust, longevity, stability, and relationships are key. The second chapter asks who. Who designs? Who do designers design for? Who is included and who is excluded? We look to the concept of *publics*, which is the foundational unit for public life. We review a range of theories regarding public versus private, the contrast between the public and the state, and the notion of plurality when it comes to public opinion, or acting in the interest of the public. And ultimately we ask how designers in public-serving organizations are working within this complexity to make necessary changes. Chapter 3 examines the role of *play* in civic life. We present play as a state of conditional freedom and explore various modalities, including games, that are particularly effective at enhancing play. We examine how play has been applied to public life, from gamification to public art, and highlight forms of play that have demonstrated effectiveness in the civic context. We look at the way that organizations, from schools to governments, have adopted games and play in their work, and how they think about potential impact. The extended case study in this chapter is a project called "Participatory Pokémon Go," which used the Pokémon Go game in Boston to involve youth in a dialogue about the use of data in the city.

Chapter 4 examines the goal of civic design—namely, *care*. Meaningful inefficiencies are spaces in which caring is nurtured. They are not just engaging systems, but systems that cultivate the conditions for people to care for the world. We examine the works of scholars who have tried to understand the role of care in social life and suggest that well beyond the caring professions (health care, teaching, social work, etc.), the phenomenon motivates and justifies actions. We use examples from the public media space of journalists who are pushing to involve publics in storytelling and, as such, reconsider the purpose of journalism in public life.

And finally, Chapter 5 looks to *practice* by providing a practical framework for designing meaningful inefficiencies within organizations. It highlights the personal and organizational challenges of doing this work and presents a resource called the Reflective Practice Guide (RPG), which is a practitioner-facilitated assessment tool for understanding and communicating the value of meaningful inefficiencies. The RPG was co-created with journalists in the United States and Europe and is being used in a variety of organizational settings. We conclude the chapter by exploring implications for civic design beyond public-serving organizations.

Civic design creates the conditions for a plurality of voices and interests to be represented, accounted for, and involved in shaping the outputs and effects of public life. It is not a genre, suite of technologies, or even set of best practices; it describes an approach to design that sits in direct opposition to the logics and actions that have perpetuated deep-seeded distrust in institutions. As media and technology dominate social life and the everyday interfacing between people and institutions, civic design, manifested through meaningful inefficiencies, presents a necessary countervalue. By focusing on and operationalizing moral obligations that undergird institutions, civic design creates objects and experiences that construct and facilitate one's connection to the world in these dark times.

[1]

INNOVATION

The history of cities is intertwined with the history of innovations. From the sewer to the subway, the elevator to the air conditioner, the automobile to the streetlight, new technologies perpetually transform how people live in cities (Marvin, 1990). The elevator made it possible for cities to grow vertically, the air conditioner made population growth possible in southern states, the streetlight extended the amount of time that streets could sustain activity. Each of these novelties, while seemingly mundane now, was a disruptive force. Each changed the way things got done and how people imagined what was possible. But the introduction of novel technology is never without struggle and negotiation. Highways made it easier for people to drive into the city; they also erased neighborhoods, specifically those occupied by poor people of color, and they starved public transit systems. Novelties come into cities like a lion, and unless there is attentiveness to how and for what reason those novelties spread, the benefits can be dangerously unequal.

Take the ride-hailing company Uber, for example. In 2010, Uber transformed the taxi industry by introducing software that coordinates drivers (in their own cars) with riders. They created a new market by allowing anyone to get into the taxi business. But of course it wasn't that smooth. There was, and remains, conflict over how the new system is integrated into existing networks, laws, and infrastructure. So while the novelty of the invention is what gets most of the attention, it has been the work of public-sector

Meaningful Inefficiencies. Eric Gordon and Gabriel Mugar, Oxford University Press (2020).
© Oxford University Press.
DOI: 10.1093/oso/9780190870140.001.0001

organizations to build the institutional infrastructure to maintain the innovation. There is a need to regulate the company's contingent labor practices; there is a need to address the influx of unmarked cars suddenly pulling over into bike lanes to pick people up; there is a need to address perpetually circulating vehicles that are exacerbating, not alleviating, traffic congestion. Building infrastructure around disruptive novelty is part of the innovation story, but it is often excluded from the narrative because it is seen as unimaginative, or the conservative force that dampens the creativity of entrepreneurs. But this is a false binary. The work of understanding what people need, what they're afraid of, and how novel inventions might empower them to imagine new possibilities is at the core of sustaining civic innovations.

This chapter is about the difference between market innovation (i.e., Steve Jobs drops the iPhone into people's lives and the world is changed) and civic innovation (i.e., novelties create the conditions for empowered publics). As public-serving organizations are transforming to meet the demands of new technologies and new social formations, more and more practitioners within them are embracing the meaningfully inefficient alternative to disruptive innovation. We focus on these civic innovators who embrace novelty, but do so to strengthen the social infrastructure in which people come together, form groups, deliberate, and advocate for themselves.

BEYOND NOVELTY

Innovation stems from the Latin *innovare*, which means to renew or change. Its first uses in the English language in the mid-1500s were to describe political revolution. It suggested a new concept that was more than simply "change," but rather the introduction of change into an established order, creating a new order in its place. The political scientist Peter Schumpeter (1939) defines innovation as "doing things differently." He distinguishes an innovation from two things with which it is often conflated. First, it is different from *invention*, which he defines as simply a discreet new scientific product or piece of technology, whereas an innovation is the act of *putting into practice* something different that generates a qualitative change in how the broader domain functions—such as a market, an organizational structure, a set of human behaviors, and so on. Second, Schumpeter stresses that the "doing things differently" of an innovation isn't just change in itself. One could change the quantity of some input in order to increase output, but that doesn't change how things are done. Rather, distinct from any *mere change*, innovation must involve "change with novelty," wherein after the introduction of the change, things are done differently.

But when operating in a civic context, where markets are composed of a diversity of publics and organizations have the responsibility to meet their varying needs, change with novelty must also include trust that novel changes will effectively serve a diversity of publics. Take, for example, one of the most notorious urban innovators of the 20th century, Robert Moses. Moses was the mastermind of mid-century American urban renewal strategies that saw the wholesale destruction of poor neighborhoods, through slum clearance, to make room for highways and business districts. In the 1950s, he sought to create streamlined urban cores, absent of poverty, through a strong narrative of innovation. This was accomplished through ostentatious acts such as slum clearance and the razing of entire neighborhoods, as well as small infrastructure changes. The sociologist Susan Leigh Star describes one of Moses's more notorious small changes. The bridges over the Grand Central Parkway in New York were to be low in height. "The reason? The bridges would then be too low for public transportation—buses—to pass under them. The result? Poor people would be effectively barred from the richer Long Island suburbs, not by policy, but by design" (p. 485). The promise of the new American city—clean, White, affluent—was the dominant vision driving the innovation of urban renewal and justified the big and small innovations Moses introduced.

There was no visible alternative to this vision at the time. The poor communities who had been displaced had also been effectively silenced and erased. It wasn't until Jane Jacobs (1969) rose to prominence, an activist from the Lower East Side of Manhattan, that the narrative shifted. She questioned the assumption that cities should be efficient machines for moving the monied class from point A to point B. She shared a vision of cities as a collection of neighborhoods, connected by vibrant sidewalks, that through their relative inefficiency, create opportunities for people to form publics. "Cities have the capability of providing something for everybody," Jacobs said, "only because, and only when, they are created by everybody" (p. 238). In many ways, Jacobs introduced the concept of civic innovation to the American public by pointing out the costs of innovating in the public realm when the result is the systematic exclusion of many of the publics who occupy it.

Decades later, the challenge and opportunity of innovation continue to guide urban policies and to shape the way public-serving organizations adapt to changing needs. While there is certainly more of a recognition of the uneven benefits of innovation, there remains an appeal within public-serving organizations for the cleanness of disruption, ported over from the private sector. Jorrit de Jong (2016), in his book about innovation and government bureaucracy, addresses the propensity for government to think in terms of client and customer, often without struggling with a more nuanced sense of citizenship. He argues that the public sector has adopted a "conception of public management informed by the purely functional transactional model borrowed from the private sector" (p. 31). The result is a clear incentive structure attached to common metrics such as number of people, services delivered, cars transported, and so on (Goldsmith & Kleiman, 2017; OECD, 2017), with very little room for the collaborative work required to build support for civic innovations.

A recent trend in public-serving organizations has been the creation of designated offices for innovations to take place, wherein there can be some flexibility with incentives. Most large cities in the

United States and Europe have opened innovation offices; national governments around the world have hired chief innovation officers; and international nongovernmental organizations (NGOs) such as Greenpeace, UNICEF, and UNDP have innovation units. The spirit behind many of these offices is to do things differently (Burstein & Black, 2014), often to transform ossified bureaucratic cultures through the incorporation of new technologies and new design thinking methodologies, which are inspired by companies like IDEO and supported by foundations like Bloomberg. On the surface, these innovation units are invested in the appearance of Silicon Valley innovation. But practically, they can take more risks because they are organizationally shielded from failure. And they also, by nature of their isolation from other parts of the organization, are invested in finding ways to integrate. They can be quite creative in how they work with other units internally and how they cultivate partners outside of the organization. Yet still this part of their story is underdeveloped.

A good example of this is Code for America. Established in 2011, the mission of this San Francisco–based nonprofit was to install coders into municipal governments to build digital tools. The "fellows," as they are called, spend weeks embedded in a government team, talking to stakeholders and shadowing employees. They then work collaboratively to identify a need, after which they head back to the San Francisco headquarters and build an app. This fellowship model was their primary mode of operation for years, and they successfully built dozens of applications for as many municipal governments. Additionally, they hold a summit every year that brings together a large swath of the government technology (or govtech) community, and they have inspired local volunteer brigades of civic coders in cities throughout the world (Schrock, 2018). Code for America is innovating, but not for the reasons most people think. Most of the apps built by their fellows fall into disuse shortly after getting built. However, a common story told by government employees that have hosted fellows is that having the fellow around motivated a department to think differently about their work. So, while the specific novelty was not sustained, being forced to think though a technology lens created opportunities to rethink institutional values and organizational frameworks. These fellows were not producing novelties, but *designing the internal conditions for novelties to be supported.*

In 2011, during the organization's first year, a Code for America fellow who was embedded in the City of Boston created an app called Adopt-a-Hydrant. This clever application allowed residents to "adopt" a fire hydrant in the city by claiming it in the app, thereby agreeing to tend to it during snowstorms to assure that it is shoveled out and accessible to the fire department. In its first year, about 100 people used the app.[1] This small experiment proved so interesting that the following year, the City of Honolulu adapted it to become Adopt-a-Siren to similarly involve the public in assuring that tsunami sirens are functional, and San Francisco created Adopt-a-Drain to assure that storm drains are clear. This is an example of a novelty catching on and becoming accepted as a way of doing business. And while what gets celebrated is the novel application of technology and the efficiency of its function, what gets ignored is the organizational process of adoption and the work involved in creating the conditions for the novelty to function.

Adopt-a-Hydrant is a story of a novelty spreading and becoming normalized across multiple organizations. But it is also a story of individual organizations changing the way they do things to accommodate a novelty. The City of Boston was motivated by the allure of invention, with the innovation remaining on the level of novelty. But Honolulu was motivated by the allure of adoption and adaptation, which, different from invention, requires the novelty to become normalized and integrated into departmental structures. Adopt-a-Siren forced conversations across departments that would otherwise not have happened. Regardless of the ultimate scale and utility of the novelty, what was truly innovative was the human coordination it motivated. The novelty was an object to think with. While it did not disrupt the business of government, it did alter, through its organizational integration, the way government conducted its business.

Here's the main point: Civic innovation is not the shiny object, but the infrastructure (both bureaucratic and public) that allows that shiny object to take on meaning beyond itself so that publics can take action. And there are a unique group of practitioners, whom we call civic designers, that are front-loading values and trust building in their efforts to innovate. In the following section, we look closely at two examples, one from outside government and the other from within: the Citizen Police Data Project in Chicago and the Student Rights App in Boston. Each is an example of a project that might have been designed purely for disruption. But instead, the process these designers went through was, from the very beginning, invested in empowering the multiple publics with something at stake.

CIVIC INNOVATION

The Citizen Police Data Project (CPDP) is a website that provides access to complaints filed against police officers in the City of Chicago. The data in CPDP comes out of decades of legal work by the project's founder, Jamie Kalven, and his collaborators to make the records publicly available. The website makes the data accessible through a graphic interface that features visualizations and summary statistics about complaints, making it easy for a user to, for example, contextualize the data by location, quantity of complaints, or individual officer.

The impetus to make these records available to the public came from Kalven's long-term experience as a reporter in the Stateway Gardens public housing project. In his time as a reporter he observed and conducted extensive interviews with residents that were regularly harassed by a group of police officers. As a reporter focusing on the realities and implications of public housing, he decided that taking a one-time policy stance on what he was witnessing would not be effective. Instead, as he put it, he sought to recruit reality from a position of marginality. By garnering input through his long-term reporting about the lived experience of residents in the Stateway Gardens public housing project, he represented the problem of police harassment by focusing on the concept of impunity, showing how officers working in spaces such as housing projects could act however they wanted.

Recognizing that officers were seldom disciplined for their infractions, Kalven worked with a network of activists and lawyers who were trying to bring more accountability to the actions of officers in the Chicago Police Department. Through extensive Freedom of Information Act (FOIA) requests and an eventual State Supreme Court ruling, Kalven and his colleagues managed to gain access to decades worth of reports about police complaints. With the content in hand, Kalven describes how he and his colleagues assumed the function in civil society of curating and making information available to the public.

While the efforts around building this database were situated within and supported by Kalven's extensive social network of reporters, activists, and legal experts, there was the unavoidable tension between his efforts and the interests of the City of Chicago and the Police Union. Recognizing this tension, Kalven notes that the allies he sought from such institutions were never those that held power, but those that operated within the institutions and saw a need for change but were not in a position to affect it. By creating the database, Kalven began to build a network of allies who were operating within police departments and saw the efforts of the database project as an outside resource they could leverage to shed light on specific abuses.

Soon after its deployment, the relationship between CPDP and the City of Chicago shifted when data from the project played an integral role in the indictment of an officer accused of shooting an unarmed citizen. While we cannot infer a causal relationship in the events that followed the indictment, the public nature of the database and the inequalities around accountability that the database lay bare during the case may have played a part in the creation of Chicago's first-ever police accountability task force.

Beyond the case, Kalven has observed that relationships with institutions of municipal and state government and the CDPD emerge as a regenerator of legitimacy, where institutions viewed with increasing distrust and suspicion regain some degree of legitimacy by publicly acknowledging the importance of CDPD in their work as public servants.

This shifting relationship with the City of Chicago demonstrates that the value of technological innovations for civic life are not dependent on any single user base or network of relationships, but are supported by a range of constituents with varying and potentially competing interests. In the case of CPDP, aligning those interests reveals that making the appeal for distributing ownership of a project is often out of the hands of the innovator and is instead reliant on tertiary actors (in this case the opinion and trust of the public) to highlight the value of the work and attract key allies.

The CPDP is an example of innovative work in the civic realm. It specifically demonstrates the extraordinary amount of effort involved in institutionalizing novelty. The standard story of innovation might begin and end with a database that was invented and shared. But Kalven's tireless work in building networks, sharing outcomes, continually listening, and transforming the technology demonstrates the texture of civic innovation and the often difficult and messy path practitioners take toward achieving it.

This idea is reinforced in a different example from within the Boston Public Schools. In 2016, youth from the Boston Student Advisory Council (BSAC) developed a mobile app that informs students about their rights if they are facing expulsion from school. The app also gives students the opportunity to file grievance reports or start conversations with administrators in the Boston Public Schools if the students think they are not getting a fair disciplinary process or find that they are the target of discrimination.

The idea for the application came from long-term school climate assessment campaigns that Youth on Board, one of the organizations that supports BSAC, has been involved with. The Dignity in Schools campaign, an independent national initiative supported by youth advocacy organizations across the United States, collects stories from students relating to disciplinary processes, their outcomes, and the impact they have on the lives of students. As one of the organizers behind the student rights app pointed out to us in an interview, the input from their long-term listening projects, which involved collecting stories around school disciplinary processes, gave them an acute understanding of the problems that needed work.

Through their long-term work of collecting stories and their ongoing conversations with an extensive network of youth advocacy practitioners, BSAC members began to explore ways in which to reduce unfavorable outcomes for students that contribute to the school-to-prison pipeline. Their primary approach to promoting equitable and just outcomes in school disciplinary procedures started with making the language around school discipline rules and codes of conduct more accessible—that is, taking a dense 80-page legal document and turning it into a user-friendly experience in the app.

While translating the dense legal language was supported by legal counsel, the grievance reporting feature required more negotiation with and buy-in from stakeholders in the Boston Public School administration and Teachers Union. Initial concerns had to do with incorporating student grievance reports into existing bureaucratic processes while also attending to concerns that students might abuse the grievance feature. Addressing these concerns around process and student use became part of a larger campaign to distribute the ownership of the app within the Boston Public School administration. Through our interviews we learned that the organizers behind the app wanted to show how it was not being positioned as a tool to be used against administrators and teachers, but was instead a tool that could improve the disciplinary process.

To demonstrate the value of the app and encourage shared ownership, BSAC engaged administration in meetings throughout the development of the app. BSAC also brought attention to the app's development by tabling in school cafeterias so as to gain student and teacher input and support. Meetings with administrators involved sharing prototypes of the app early and often for user testing and feedback, as well as extensive conversations about how the app and the grievance reporting feature would fit into and modify existing school disciplinary processes. At one point in the process, the Boston Public Schools took over user testing and the development of the app as a way to ensure it would work with their existing information technology systems.

From making the language around school conduct more accessible to giving students more agency, the work of changing how schools engage in disciplinary procedures was not a matter of disrupting business as usual through a novel technology. Rather, it was a matter of carefully and strategically building the technology with all stakeholders, from IT administrators to students.

Such stories contrast the narrative of disruptive innovation, where new technologies are thrown into market without input from potentially contentious stakeholders, leaving those who don't fit new models scrambling to figure out how they will survive. What cases like the student Boston Rights App and the Citizen Police Data Project show is that this Darwinian approach to deploying new technologies and services does not work for the civic sector. For example, if the developers of the app had not brought administrators and teachers into the discussion of creating the app early on, the fears of student misuse would have likely undone the project.

The cases of the Citizen Police Data Project and the Boston Student Rights App demonstrate the required effort necessary for the transformation of novelties into accepted features of an organization's day-to-day operation, highlighting how such trajectories of adoption contrast with narratives of disruptive innovation. Each case involved the invention of a digital tool, in the case of CPDP, from outside a large organization, and in the case of Boston Student Rights App, from inside. But in both cases, the innovation was not simply inventing to disrupt an existing model of practice, but in supporting groups of people to cultivate new models of practice.

This is the actual work of civic innovation. Volumes have been written about Silicon Valley and innovation culture, largely focused on the individual achievements of the (mostly) men who "move fast and break things" (Taplin, 2017). And an equal amount of attention has been given to all the methodologies devised to cultivate this sort of creativity and invention. From creativity labs, to design thinking teams, in the corporate world to innovation teams in the government space, there is no shortage of approaches to creating more novelties and helping them to find markets. But what of that other dimension of innovation we have been discussing, where novelties inspire organizations to connect in different ways, and where they become objects for publics to think with, to build trust with, and to inspire new sorts of actions in public life?

FOUR ACTIVITIES OF CIVIC DESIGN

Through our research with public-serving organizations, we have identified four distinct activities in which civic designers engage in order to innovate (see Table 1.1).

The first activity is network building. Civic designers place value on informal gathering spaces that bypass some of the strictures of formal meetings or input sessions. Such spaces, including community centers or social media interactions, support encounters between stakeholders and allow people to identify critical mass around local issues as well as explore possible approaches for taking on particular challenges. These sorts of encounters, whether online or offline, build networks that further enable opportunities for sharing experiences and knowledge. In our conversations with practitioners, such informal spaces were described as essential to their work. Take the Anti-Eviction Mapping project, for example. It is an organization in Oakland, California, committed to engaging

Table 1.1 NAMES AND DEFINITIONS OF ACTIVITIES ASSOCIATED
WITH CIVIC INNOVATION

Activity	Definition
Network building	The act of convening either in person or online for the purpose of social connectivity.
Holding space for discussion	Assuring that there is time and space for discussion that makes room for multiple viewpoints and is tolerant of dissent.
Distributing ownership	The designer or convener takes time to build capacity of all stakeholders to reproduce or modify designed activities.
Persistent input	Inputs into products or processes from stakeholders continue beyond initial release or implementation.

Source: Gordon and Mugar (2018).

housing-vulnerable people in policy solutions. According to the founding director, Erin McElroy:

Since doing this work, my own sense of community in the Bay Area has shifted dramatically, and it's really nice to know—to have good relationships with people in different neighborhood coalitions. Whether it's a housing clinic or a legal organization, it's nice to know that I can email or call or, you know, show up to somebody's office and people know who I am and who the Mapping Project is. It's nice to kind of feel that we're not single-handedly having to do anything but that we're one of many groups doing a lot of things, and that there's some sort of a network, and people generally understand where we are. (personal communication, 2017)

The second activity is "holding space." This describes all the work involved in assuring that everyone is at the table and that the table is actually set for everyone. When describing the work they do, people discussed the work involved in holding regular meetings and workshops, where coming up with the guest list or designing the physical space itself was a huge part of the effort.

City Bureau, a community journalism organization working in Chicago's South Side, hosts a Public Newsroom, which is a weekly gathering at their offices where journalists and members of the public discuss local issues, share information about emerging stories, and support residents in conducting their own reporting efforts. The decision to do this on a weekly basis came as a measure to counteract a long-standing distrust of journalists in the South Side. This distrust is due to the standard relationship journalists have with communities, where they show up to do a story and disappear when the story is done. So the innovation is not the public newsroom itself, but the work involved in socializing it.

Andrea Hart, City Bureau's Community Director, notes that there is an uptick in applications to participate in the program, something she points to as evidence of growing trust in City Bureau. "I don't know if we're necessarily part of the fabric of the community yet," Andrea Hart told us. "I think that's going to take a long time, but I definitely think it's a mutually beneficial relationship where we try to assess needs around information and issues that they care about, and then try to go back and do reporting or do some sort of project that then we just bring back and have dialogue around and help inform folks to make better decisions" (personal communication, 2017).

The third activity, "distributing ownership," involves all the work of assuring that people feel a sense of ownership in the innovation. Practitioners describe clear pathways to participation, actively encouraging a power dynamic where stakeholders take the reins of the practice, or sharing knowledge and encouraging appropriation and repurposing of practice.

The Gray Area Foundation, an organization that supports civic art projects in the Bay Area, described the practice of distributing ownership as the neighborhood taking care of art projects after the initial innovation is complete. This requires artists to attend neighborhood meetings and build consensus around the objectives of the work so that there is clear value for everyone involved, even if they're not motivated by art. As their Director of Education points out, "We have a strong focus on creating scaffolding and open-source structures so that other cities can pick up the ball" (personal communication, 2017). In Chicago's City Bureau, distributing the ownership of their innovation is carried out when project alumni go on to do their own work and develop their own professional networks. They are then able to activate that network when they work in the same area. They do this through online tools such as Facebook Groups and Slack, where they maintain steady communication with alumni, keeping them engaged with the organization and with each other.

The need to distribute ownership was central to CUT Group (Citizen User Testing Group), an organization that engages Chicago residents in the testing of civic technology. The people involved in this initiative invite user testers in their network to become proctors, coordinating and overseeing user testing sessions. By inviting active members of their network to take on leadership roles, CUT Group increases the capacity of their services while also broadening the number of people in their community that have the skills needed to conduct citizen user testing work. By promoting testers to leadership positions, CUT Group not only sets itself up for long-term success, but also bolsters its capacity to take on more work.

The final activity is "persistent input." Practitioners understand the context of their issues by not simply asking people what they think, but doing so from a position of stability, continuity, and trust: asking once, and then being in the same place to ask again. This persistence is reflected in long-term relationships between practitioners and the communities or larger organizations they work in.

Jamie Kalven, of the CPDP, notes that while he didn't know how to solve the broader problem of police brutality, he knew that there was a problem, and he wanted to "recruit reality" as a way to highlight a problem, even if he didn't know what would fix it. "I'm not a policy guy," says Kalven. "I don't know what to do with public housing, but I'm standing here. And I was standing here yesterday and I'll be standing here tomorrow . . . I know this about these conditions. So it was kind of recruiting reality to our ends" (personal communication, 2018).

Kalven recognizes the importance of persistence—that understanding problems is not simply a matter of asking people what they think, but doing so over time from a position of trust. He sees this as the fundamental distinction between standard institutional practice, mired in bureaucracy and lacking in relationships, and emerging institutions built on new value propositions. For novelties to institutionalize, the various stakeholders of that novelty need to trust that it is not just a flash in the pan, that their investment of time now will lead to something in the future.

The four activities—network building, holding space, distributing ownership, and persistent input—describe the work involved in civic innovation. For this group of practitioners, there is a constant negotiation between the values inherent in the technology and the values guiding the institution. Jamie Kalven introduced a novelty to the police department, but he needed to build trust among the communities he was representing and within the police department. He engaged in a copious amount of effort to not only win support for the effort, but also to assure that ownership was shared across a range of stakeholders. He wasn't just looking for a market for his novelty; he was assuring that the values of the novelty aligned with the multiple publics that would form its constituency. Likewise, the Boston Student Rights App was conceived as a novelty, but the people involved in its creation knew that the real work was in building a shared sense of value as a means of assuring the novelty's persistence. This is a great example of a tool, still in use at the time of this writing, that needed to have a robust coalition of internal stakeholders in order for the project to form and retain its value.

In highlighting the four activities of civic design, there are important similarities to other design practices and value systems, in particular those that have shaped approaches to design that navigate multistakeholder collaboration across asymmetrical power relations. For example, participatory design is an approach to bringing end users of services and products into the design process such that they have equal weight in the decisions being made. This commitment to creating greater symmetry in the design process has its roots in the Scandinavian workplace democracy movement (Bjerknes, Ehn, & Bratteteig, 1987), where labor organizers brought management and workers together to create more equitable working conditions. Because there are no dominant theories or practices for participatory design, there are a multitude of contexts to which it is applied as well as approaches to do this work. As Muller and Druin (2002) describe, participatory design appeals to researchers and practitioners who see that creating processes for increasing representation of diverse knowledge and backgrounds can make for better products and services. The theory and practice of *design justice* offers another layer of urgency to our work by highlighting the ways in which oppressive power dynamics can be replicated at scale through emerging technology. Design justice encourages both critical reflection among designers about these power relationships and, like participatory design, pushes for a centering of marginalized voices in the design of products and services that can perpetuate harmful social institutions (Costanza-Chock, 2019). Lastly, civic design resonates with what design theorist Chris Le Dantec calls social design (2016). In his study of a participatory design process used to develop a tool to support staff at a homeless center and their constituents, Le Dantec describes how relationships in a participatory design process are reified through social and technical means. These relationships that emerge around the design process create an infrastructure for the circulation of ideas and interests which Le Dantec describes as design publics.

Like Le Dantec, we are interested in the formation of publics in the design process. But we are interested in shedding light on those processes where the design of publics is a more intentional objective, one that sits equally alongside the goal of designing an artifact or service. Furthermore, we see this as a practice that exists not just at the level of the design practitioner, but within organizations themselves. For the most part, the people we talk about in this book are not professional designers. They are practitioners within organizations that are socializing and institutionalizing novel inventions for the purpose of supporting trusting publics.

AMBIGUITY

The four activities described earlier are the practices of the civic designer. These designers embrace the intentionality of the activities, but they also are comfortable with open-endedness and unforeseen possibilities. For example, when Jamie Kalven set out to create the CPDP, he knew he was addressing an important need. How and by whom this would be adopted and the type of change it would affect were by no means a foregone conclusion. The impact that the CPDP has had on the institution of law enforcement in Chicago was a product of the provocation to bring multiple stakeholders together into conversation around the novelty of the database. Its value to various institutions emerged through the relationships that formed over time around the database, with the tool taking on distinct roles for different individual and organizational needs. The tension between the certainty of a product's functionality and the uncertainty of its use was very productive. It generated an ambiguity that allowed for relationships to form as people were drawn together by their questioning.

Everything we are suggesting runs counter to accepted norms of market innovation, which tends to focus on the quick, transactional, and temporary. Civic innovation is slow, relational, and persistent.

In the context of government work, de Jong (2016) makes the point that innovation in the public sector needs to be understood from the perspective of the client *and* the public. "The fact that the public often is not articulate about its values, not coherent and consistent in its preferences, and not organized to act on its interests makes it both harder and more important to look at bureaucratic dysfunction from a public perspective" (p. 50). He is describing a condition of ambiguity that is desirable when seeking change within public-serving bureaucracies. This state of ambiguity is a required condition of civic innovation, where designers are working with not just one public, but multiple publics. Civic innovation includes the explicit articulation of the value of a novelty so that it might find its role both within the bureaucracy and within and between the publics it serves. Ambiguity in this case is not a bad thing; it is necessary. But for most practitioners, who are operating with deadlines and deliverables, it can be a deeply uncomfortable place to be. Design firms like IDEO have called out ambiguity as a state that designers need to embrace. Being comfortable with ambiguity mitigates the tendency for designers or anyone in a consulting position to feel like they need an answer at every moment. Embracing ambiguity gives designers permission to dive deep into understanding the nuance of a particular context, bringing their clients along on the exploration as they speak with potential users and test out early ideas. The role of ambiguity in a civic design setting is indeed quite similar; however, because civic design starts and ends with publics means that the boundaries defining needs and interests can be far messier and nebulous than what a client relationship might contend with. To factor multiple publics into decision-making processes, to assure trust and ownership across multiple stakeholders in novelty, and to create spaces for vulnerability and play among competing interests requires that practitioners and organizations tolerate a certain level of ambiguity. And for innovation to be done right, well beyond simply tolerating it, ambiguity must actually be incorporated into the design process. As we highlight in case studies both in this chapter and throughout the book, the open-ended exploration of needs and interests of multiple publics provides the required flexibility to create valuable product and experience. The alternative to being open ended in these cases would mean accepting the default values of novelty and bypassing the complexities and contradictions of working with multiple publics.

Ambiguity cannot be a permanent state. Indeed, there is no better way to squander trust with publics than to accomplish nothing. But as we have been describing, the civic designer self-consciously constructs a space for play before arriving at specific objects or outcomes (see Chapter 3). This space for play allows the designer and the stakeholders to reflect upon the moral and ethical dimensions of the design's mechanisms before they disappear into the mundane (Latour, 1998). From the perspective of IDEO, this is just good design process, whether it's a smartphone or a coffee maker. But civic design begins and ends in a different place than the typical design process. It always starts with the negotiation of a plurality of publics and always ends with that plurality caring about shared outcomes (see Chapter 4).

Take 311 as an example. Many municipal governments around the world have created or incorporated tools that enable citizens to report urban problems, such as potholes, graffiti, outed streetlights, and so on. The technology quickly found a use case within government operations and is now a regular part of doing business in many cities (O'Brien, 2018). However, in this case of a "successful civic tech," few groups have challenged the value proposition of such a tool. Who is best served by the technology? Who is excluded? What kinds of things can one report? Who gets to decide? Where do the inequities exist in reporting, and what is the responsibility of government to address them? These questions are difficult, and they would potentially run counter to the goal of increasing government efficiency in solving everyday urban problems. So, in the majority of use cases, the ambiguity of 311 has been quickly dispelled for clarity of function.

Simone De Beauvoir writes in *The Ethics of Ambiguity* (2011) that it is easy, when faced with ambiguity, to "follow the line of least resistance" into unambiguity, whatever that may be (p. 220). Despite the many bureaucratic obstacles to getting things done in large organizations, there is a strong push to either get novelties adopted or to quickly put them into disuse. It is precisely this ease that the civic designers we discuss throughout this book are resisting—they are assuring that what was at once an accepted matter of course may become difficult. Of course, there are many difficulties in bureaucracies, but it is often the case that the default difficulty is actually the easiest thing. It is not generative, productive, or expansive; it is, as we describe in the introduction, a *mere* inefficiency. Civic innovation comes with a responsibility to embrace the more difficult form of difficulty, one that inserts ambiguity, but always with intentionality.

To be clear, a difficulty is not useful when things are already difficult. When trash piles up on the streets of Cairo because of a breakdown in infrastructure and political conflicts (Hessler, 2014), adding difficulties is precisely the opposite of what is needed. But when a local newspaper continues to ignore the stories of poor residents because it is too difficult to build trust with those residents, then difficulties are precisely what are needed. When complacency and habit begin to creep into those who see themselves as "doing innovation," it seems incumbent upon designers to learn from Tom Sawyer, who, upset that a particularly serious situation is no fun for him because the solution is too easy, says to his friend: "Why drat it, Huck, it's the stupidest arrangement I ever see. You got to invent *all* the difficulties" (2017, p. 103).

The object of civic and market-based innovation could be the same thing. The difference is what gets highlighted when its story is told. The CPDB is a publicly accessible database. Finding the data to populate the database and building a user-friendly interface, are both interesting and important novelties. One version of the innovation story ends there. But the cultivation of public dialogue, relationships built with impacted communities, connections to public-sector actors, comprise the *civic* innovation story. That said, most public-serving organizations highlight the former story.

The objective of this book is to celebrate all those innovators within these organizations that are highlighting the latter.

CONCLUSION

The cases in this chapter offer insight into what innovation can look like in public-serving organizations. Creating a novel object or experience that is desirable is, of course, an objective that still counts for civic designers. But equal to the emphasis of creating something useful and delightful is the emphasis on creating the conditions for which people can have conversations around the new product or service. This is not conversation for conversation's sake, playing into stereotypes of bureaucratic inefficiencies, nor is it uniquely focused on the features and characteristics of what is being designed as we would see in participatory design; rather, part of what is novel is the way in which civic designers have created an opportunity for people to come together, share ideas and concerns, and build trusting relationships (see Table 1.2).

Table 1.2 DIFFERENCES BETWEEN MARKET AND CIVIC MODELS OF INNOVATION

Models of Innovation	
Market Innovation	*Civic Innovation*
Markets	Publics (see Chapter 2)
Consumption	Play (see Chapter 3)
Disruption	Care (see Chapter 4)
Transactional	Relational (see Chapter 4)

As we have shown in this chapter, civic innovation suggests a series of distinctions that set the task of the civic designer apart from a designer working to create a novelty that responds to market desirability and is inspired by goals of efficiency. First is the motivation for innovation. Where design inspired by markets looks for a discrete product or service that responds to a specific need, design for publics requires creating a space that will help the designer uncover and articulate a need. The examples we provide demonstrate how designers start with a provocation for a product that is generative in the way it brings people together to engage in conversations about issues that matter to them. In doing this, civic designers create the conditions for people to come together and build connections.

By focusing on creating opportunities for people to engage in discourse, the civic designer pursues adoption or uptake through a relational rather than transactional process. From City Bureau to the Boston Student Rights App, adoption is not a phenomenon wherein the value proposition of the object or service is self-contained. The decision to buy a product from a large corporation might be motivated by decades of built-up trust in their product line. In such cases, the consumer does not have much more to rely on other than product reviews or friends and family who own the product. A consumer is not in a position to have a conversation with employees of the business to get a behind-the-scenes understanding of how the product was made. Trust in this case is built into the transaction of acquiring the object. In the context of civic design, trust is relational, where stakeholders are empowered to engage in dialogue with each other and with the long-term success of innovation being a product of ongoing relationships. The Boston Student Rights App would not have been deployed had it not been for extensive stakeholder meetings and co-design sessions. City Bureau would not have had success had it not been so open and inviting of input from its surrounding neighborhood. Building trust and gaining adoption through reputation are indeed important objectives for any designer, but the transactional nature of relying on reputation alone, as we have shown and will continue to show throughout this book, does not work for civic contexts.

Civic innovation is a long game. It is not focused on radical change but is instead calibrated to incremental change. The two cases we offer in this chapter are a testament to this. Both the app and the database could have easily been framed as adversarial to a specific group of stakeholders. At first glance, we might frame the app as helping students fight back against school administrators who are doing an injustice to students. Similarly, the database could have been framed as a technology used to fight back against police abuse. Instead, both cases were deployed in settings that resulted in a collaborative experience between potentially adversarial stakeholders. Where Uber, the example we used to start this chapter, is locked in a seemingly eternal struggle with people whose livelihoods depend on livery services, the app and the database created common ground and surfaced shared interests that allowed people to come together.

Civic innovation is the product of civic designers. These practitioners, while not yet organized, share common methods and values. In the chapters that follow, we explore what this looks like in a range of contexts and organizations: from the shift from markets to publics, to understanding how consumption is reoriented to play, how the allure of disruption is transformed into opportunities to care, and how models of transaction are reoriented to relation.

PUBLICS

Many-to-many communication is characteristic of digital networks. Information often spreads through personal communication that gets reproduced and transformed into other people's personal communication. One's perceived sense of public life can be manufactured entirely through personal correspondences online. And yet there are a myriad of public-serving organizations, specifically in the government and news sectors, whose task it is to mediate public life. Benedict Anderson (1983) points out that in the 19th century, newspapers were core to one's ability to imagine a community of people beyond their direct experience. Newspapers constructed a narrative into which individuals were able to map themselves. This is what it means to be British, or German, or a part of a particular city. In a contemporary networked society, however, those narratives are far more complex, and organizations no longer have the same level of authorial control. *The New York Times* no longer has significant influence over what it means to be a New Yorker. Public identity is crafted through countless archived conversations, shared artifacts, and captured moments. So then, how does one's sense of public life coalesce? And what is the role of public-serving organizations in mediating that perception?

This chapter is about publics—how they form, whom they include or exclude, and how they function to get things done in the world. *The public* is not just an undifferentiated mass of people; it is certainly not just a dormant market waiting to be activated. It is

Meaningful Inefficiencies: Designing for Public Value in an Age of Digital Expediency. Eric Gordon and Gabriel Mugar, Oxford University Press (2020). © Oxford University Press. DOI: 10.1093/acprof:oso/9780190870140.003.0003

multiple. It is *publics*, comprised of a plurality of interests and realities that are defined as much by their differences as by their similarities. As we describe in the previous chapter, innovation in the civic realm, or deliberate changes in the quality and function of public life, requires an understanding and commitment to the various ways in which people make sense of, and find meaning in, being in public. Who gets to speak for whom? How is trust in representation cultivated? Civic design is not focused on individual user needs, but instead, how that user's connection to a public guides his or her experience. What sets civic design apart is a focus on publics, not users. Understanding how and why publics form, and how organizations actively contribute or resist these formations, is essential in the practice of civic design.

When the news broke in 2017 that the movie mogul Harvey Weinstein was sexually assaulting women, the public outcry was extensive—not so much in reaction to the deeds of one powerful man, but in reaction to an industry and indeed a public that remained silent about the abuses of women. The powerful #MeToo campaign emerged as an effort to draw attention to a commonality of experiences for many women, and to point out that sexual assault or misconduct is not a rare occurrence, but the norm. This "casting couch" culture was recognized and generally ignored as a private matter, until #MeToo brought it into stark relief as a matter of public concern. This is an example of a rapid and radical transformation of the definition of the public. The campaign for Black Lives, or #BlackLivesMatter, is another example of a popular push toward inclusion in dominant definitions of public in the United States.

Publics are not some static assortment of places and topics; they are constantly shifting spaces comprised of a multitude of platforms, genres, and modalities. The examples of #MeToo and #BlackLivesMatter are popular campaigns that generated quick transformation through struggle, but publics come in many sizes, and their formation and transformation are not limited to popular campaigns and scandals. On the level of neighborhoods, ethnic communities, interest groups, or business associations, publics are expanding, intersecting, diverging, and transforming, which makes them an incredibly difficult medium to work with.

What's more, in this ever-shifting model of publics and their interests, the question quickly arises: Who gets to speak for publics? The political philosopher John Dewey, in his book *The Public and Its Problems* (2012), struggles with the very notion of representation. How is that one person or even one group can speak for others? Of course, what is the alternative? Surely if every individual spoke for herself, it would be utter chaos. But this representative model requires a leap of faith, trusting that there are effective processes in place for an individual to capture and speak to the interests of many before an authoritative body or even to other publics. Public trust in civic institutions can, from Dewey's perspective, be understood as belief in a system and process of representation. If current processes are not communicating the interests of a public to a governing body, then distrust in civic institutions will arise.

Consider the example of the controversy over Boston Public School start times that we discussed in the Introduction. Think about the picture of the man holding a sign that says "Families Over Algorithms" and consider this from the perspective of Dewey's investigation into how one represents the issues and concerns of many. For the man holding the sign and the many others in the room protesting the decision, the concern was that, for a civic matter such as public education, the interests of the people were not captured in such a way that represented the interests of parents. Certainly an argument could be made that the complexity of bus routes and their optimization against commuter traffic in Boston was a push for representing the interests of many Bostonians; however, the parents who had a direct stake in this decision felt that the line of communication between their shared interests and the people who make decisions was cut off, excluding their interests in favor of a model that could efficiently contend with what was being framed as a complex systems issue.

The interests that constitute the public of Boston Public School parents is not represented early on, favoring instead an approach that appeals to a set of interests defined by engineers that create the data processing technology through which existing traffic patterns are understood and optimal pathways are discovered. While the intent to find the best bus route through cutting-edge data analytics is not nefarious or ill intentioned, it does present itself as a recurring problem in the history of public life: that of who (or what) gets to speak and whose ideas and interests are represented in proposed solutions.

Civic design is design for and with publics. It is public centered, not human centered. In this chapter, we provide a quick history of publics in Western democracy, explain how the concept is in perpetual transformation, and provide examples and analyses of people designing meaningful inefficiencies to create, nurture, and sustain publics.

THE EMERGENCE OF THE PUBLIC

The concept of a public is not a given in the history of democratic thought. While a body of people holding a state apparatus accountable to their interest has roots in ancient Greece, the project of a government for the people by the people is still a new phenomenon relative to the totality of Western history. To understand the fragile nature of democracy is to understand the existence of a public in the first place. What are the conditions that allow for people to come together and share concerns while also being in a position to express those interests and make demands on an authority entrusted with defining matters that shape their lives?

To answer the question of what makes publics possible, we begin by drawing contrast to moments where the public did not exist. The most obvious example is European monarchies in the Middle Ages, where there was what Craig Calhoun describes as "representative publicity," in which a King and his court were the public, with his interests dictating the governance of his country (1991, p. 7).

In this context, the King is not accountable to his subjects. His subjects have no vista from which to monitor the King nor do they have an opportunity to convey their concerns. Similar to how Seyla Benhabib describes totalitarian governments, "there is no spatial topology; it is like an iron band compressing people increasingly together until they are formed into one" (1991, p. 77).

In 18th-century northern Europe, the decompression of this band of totalitarianism and the emergence of a space between the people and the King was aided by the rise of print media and its use in long-distance trade. Along with the trade of goods came the circulation of information about prices, demand of goods, and general news about what was happening in the different cities (Calhoun, 1992). The circulation of news along the trade routes was coupled with what Jürgen Habermas (1989) describes as a growing social scene in 18th-century salons and coffeehouses, where news was discussed among merchants who shared a mutual concern about the operations and successes of their business.

With growing trade, the interests of local economies extended across national territories, requiring kingdoms to create the means of protecting and supporting the needs of a growing merchant class. How people came together to discuss and make demands on the ways in which their interests were being supported by another entity is what created the distance between the people and the authority of a king, allowing for a power relationship to emerge in which the needs of the people became important. As Calhoun (1990) writes:

> Civil society came into existence as the corollary of a depersonalized state authority. It became possible to recognize society in the relationships and organizations created for sustaining life and to bring these into public relevance by bringing them forward as interests for a public discussion. (pp. 8–9)

What was once a world in which all life and decisions were self-contained within the court of kings became one in which people outside of the governing bodies were able to make demands. The distance that emerged between the interests of the people and the authority of the court fundamentally changed the dynamic of governance. Habermas famously describes this emergent gap as the *public sphere*, or space in between private citizens and the state where private citizens circulate ideas, engage in discourse, and lift up their collective individual concerns to public relevance.

The emergence of a public in 16th-century Europe represents a shift in the power relationship between those governing and the governed. Where there was once no space for discourse and building alliances to make demands on the ruling power, the emergence of a public sphere where people were better equipped to make demands shifted the purpose of what a governing body or state exists to do. Where a king was not accountable to his subjects, the head of a state in a democracy exists only to represent and act on the interests of citizens. In a democracy, the purpose and function of government are defined by the public. The direction and vision of government policies should only be defined by the interests that have been expressed by the grouping of private citizens that form a public.

The purpose of a public, then, is to act as a mediator between the state and the society of private individuals, articulating and making known issues of shared importance and holding the state accountable to addressing these concerns (Fraser, 1990). This work of holding the state accountable also works to preserve a zone of autonomy for private citizens, preserving their agency against the domination of the state (Calhoun, 1992).

The work of a public holding the state accountable has been described as a mode of social coordination (Drew, 2017). A pool of private citizens coming together to address their concerns to the state requires people to mobilize and interact with one another, attend meetings, coordinate protests, and write down demands. In the case of the new merchant class of 16th-century northern Europe, the coordination worked to align conversations around questions of commodities exchange and the rules and conditions that supported the trade routes. The actions that followed involved debating the state around the rules and norms that defined the conditions of trade (Calhoun, 1992).

The public sphere emerges as a mechanism for negotiating collective interest and as a means of influencing state power. But from the moment of its emergence, this mechanism has continuously transformed along with social, economic, and political tides. How individuals come to be part of the public, or a public, is not uniform. As Warner (2002) states, a public is an "ongoing space of encounters for discourse" (p. 90). In the following section, we provide a basic schematic of how publics work and how people negotiate their boundaries.

HOW PUBLICS WORK

Publics involve people sharing and discussing issues that are important to them. Habermas locates the birth of the public with the advent of the printing press and the Protestant Reformation in the 16th century. The experience of religion prior to the printing press was dependent on word of mouth, where people would only hear what was presented by the church and then communicate with each other. With the rise of literacy along with the printing press and the push for a more personal experience with religion, people were poised to engage critically with ideas rather than believe only what was spoken to them.

This was coupled with a growing market economy that created interdependencies that extended far beyond the boundaries of a village. In his account of how the merchant class of the Hanseatic league came to create a critical mass of influence that existed outside of the courts, Habermas points to a combination of the printing press that allowed for a high volume of written content to travel throughout northern Europe along the trading routes. In the cities where this news would reach, people with business interests would convene to share and discuss news and, in particular, discuss how various authorities through northern Europe were supporting the interests of their trade.

Supporting the circulation of content were the 3,000 coffee shops in London during the early 18th century, each with a core of patrons. Here, patrons would discuss the news as well as read journals of opinion. With many of the patrons at these over 3,000 coffee shops reading similar journals, people were connected through the content they consumed across the city, the country, and the entire network of trade. While people were able to see their shared interests through conversations with fellow coffee shop patrons, there was also what Benedict Anderson (1983) describes as "imagined communities," where individuals felt a connection beyond their immediate periphery to the people in other cities who were reading the same content.

Calhoun (1992) suggests that this literary public sphere was essential to supporting what he calls zones of autonomy, or the ability of people to circulate ideas among themselves that were distinct from any institution related to church and state.

> The literary public sphere helped to develop the distinctively modern idea of culture as an autonomous realm: Inasmuch as culture became a commodity and thus finally evolved into "culture" in the specific sense (as something that pretends to exist merely for its own sake), it was claimed as the ready topic of a discussion through which an audience-oriented subjectivity communicated with itself. (p. 12)

These zones of autonomy can be self-sustaining. Consider art collectives such as the Bauhaus or the Futurists that would circulate their own literature, often in their own journals, as a means of creating influence and perpetuating a worldview. Today, it is possible to consider Facebook groups or sub-Reddits as zones of autonomy. Any place where people are able to freely share ideas within defined boundaries can be constructive of a public identity. Ray Oldenburg, in his book *The Great Good Place* (1999), is interested in the physicality of such places, dubbing them third places. Beyond work and home, third places are anchors of community life that facilitate broad and creative social interaction. Third places, in that they are distinct from traditional sites of domesticity and workforce labor (they can include coffee houses, parks, and online forums), are able to support the free circulation of ideas.

Of course, these locations are never pure or absent of power differentials. They can be sites of exclusion, silencing, even abuse. And they can be idealized and commodified. Starbucks founder and one-time presidential candidate Howard Schultz (1999) mass-marketed cafe culture as a third place, and as a result he was able to substantially raise the price of coffee. The innovation of Starbucks is certainly not the coffee; it's the third place in which one can drink it. What's happening here is the privatization of the public sphere that puts the act of consumption at the center of the formation and perception of publics.

Habermas (1989) warns that the normalization of publics as consumable spectacle can have wide-ranging political effects. There is the state and the private individual, in between which is public discourse, where private individuals represent themselves before the state. What Habermas sees taking place in the 20th century is the refeudalization of the public sphere, where the state and private entities begin to take on tightly knit relationships that capture the interest of the few over the interests of the many. This dynamic between the state and the private sphere is a product of special interest groups and lobbyists capturing the attention of legislators. The effect of this dynamic is that, behind closed doors, legislators look for support rather than engage with the public in rational critical debate about the interests that need to be represented (Calhoun, 1992). With their connections to special interest groups in line, legislators, like the kings before the public sphere, represent interests before the people, rather than for them. It is this representation before the people that creates the *public as spectacle*, rather than the public as discursive body. According to Habermas:

> The process of the politically relevant exercise and equilibra-
> tion of power now takes place directly between the private

bureaucracies, special-interests associations, parties, and public administration. The public as such is included only sporadically in this circuit of power. (1989, p. 176)

In the public as spectacle, the work of a king or a legislator is to derive power through the way they present themselves. Habermas believes that during the public sphere of the 18th century, the theater of power was replaced by rational debate, only to return again with the greater intersection of private and state interests (Calhoun, 1992; Habermas, 1989). That Starbucks finds a mass market for third places, and that publics emerge in private social media platforms like *Facebook*, points to a blurring of lines between public and private and the normalization of publics as consumable spectacle.

In this political context, private individuals are compelled to consume personas, not ideas. As Warner points out, people don't see their concerns being represented; rather, they see themselves reflected in a mass media icon (Warner, 2002). This is true with "friends" on Facebook as well as elected representatives. This idea of connecting with a public political figure rather than the issues they represent has played out quite explicitly in recent elections in the United States, where public reaction to presidential candidates has come down to describing them as relatable and someone you might want to have a beer with. This perceived relatability is the very effect that the deeply asymmetrical public as spectacle needs to create, manufacturing a sense of reciprocity so that private citizens feel like they are getting something in return (Warner, 2002).

So what does all of this mean for the civic designer? What does it mean for that person who is looking to innovate how publics are recognized and supported by their organization? If there is no such thing as "*the* public" or "*the* public good," then how does one define goals? If publics are always multiple and malleable, then where does one begin? As we have discussed, publics (whether singular or multiple) are what keep governments accountable to the interest of the people, act as spaces for members of society to discuss and coordinate their interests, and help reveal new interests that are not yet part of the public consciousness. This is the object of civic design. How designers define objectives, boundaries, and connections, and balance the value of constraints with the dangers of exclusion, is what sets civic designers apart from others (DiSalvo, 2012).

PUBLIC VERSUS PRIVATE

What topics are people allowed to talk about in public discourse, and what topics are inadmissible? With the rise but albeit slow acceptance of identity politics in American society, the question might seem backward and unenlightened. But this is the question that sits at the core of how societies grappled with a pluralism of issues that could potentially overwhelm public discourse. From the Greek polis described by Aristotle to the public sphere of emerging republics in the 18th century, the contours of what could and couldn't be discussed rested on a number of criteria; principal among them was ensuring that the topics discussed were neutral, or matters of the common or public good, rather than matters of private concern (Benhabib, 1992; Fraser, 1992).

Building on the model of the polis in ancient Greece, Hannah Arendt (1998) unpacks what she describes as this distinction between public and private matters, relegating "certain types of activity like work and labor, and by extension all issues of economics and technology, to the public realm" (Benhabib, 1992, p. 80). Benhabib further elaborates on this Arendtian classification noting that the public-private distinction places such work as domestic labor—for example, housework, reproduction, and child and elderly care—as private matters that had seemingly natural or immutable aspects that do not need to be explored in public discourse (Benhabib, 1992).

This act of bracketing out what could and could not be discussed was done with the intent of achieving some idealized model of discourse that cut through personal interest and existed only on the plane of common ground. In his study of the rise of the bourgeois public sphere in 18th-century Europe, Habermas describes modes of discourse that bracketed out class distinctions such that people could engage in rational and objective conversation on neutral ground.

While the logics that defined what was public and private in the context of the polis trend toward one end of the spectrum, it serves to represent a model of bracketing that leaves little room for what can be discussed. Later models of public conversation would include matters of economics and trade; however, the question of what could be considered public and private functioned primarily as a means for the production of class distinction (Bourdieu, 1990) that in turn legitimized the oppression of genders and races (Benhabib, 1992; Fraser, 1992). For example, Benhabib points out that the use of the public-private binary has been taken to task by feminist theorists, showing that "traditional modes of drawing this distinction have been part of a discourse of domination that legitimizes women's oppression and exploitation in the private realm" (1992, p. 93).

This model of bracketing conversation around what does and does not count as a public issue helped to establish publics as exclusive places for participation, reinforcing notions of class and gender distinctions in society. The historical explanation of the emergence of the public sphere offered by Habermas has been extensively criticized for this reason. Habermas's critics highlight that he offers a narrow definition of what interests can and cannot be shared and that those interests typically benefit a specific class of people, namely wealthy merchants who were men, educated, and propertied. The justification of property ownership and education was meant to provide exclusive participation rights to people who were "autonomous" and capable of rational-critical discourse (Calhoun, 1992).

Nancy Fraser (1992) points to a wide range of scholars that have examined the role of gender in defining participation in public life. For example, Fraser points to the work of Joan Landes, who describes the French public sphere of the 18th century that was constructed as an opposition to the salon culture that welcomed women. Styles of discourse that characterized a more "virtuous" and "manly" tone were favored, leading to an eventual formal exclusion of women from public discourse.

In the Greek polis, public discourse was described as a leisurely activity, where men would engage in debate about civic matters with the aim of gaining public prominence as skilled orators. The polis was described as egalitarian and morally homogenous (Benhabib, 1992), making it possible to assume a shared sense of common ground before arguments began. The egalitarian, morally homogeneous, and leisurely conditions of the polis were, however, a product of all the people who were excluded—including women, slaves, children, laborers, and so on—engaging in the work of the private realm that made time for men to participate in public discourse.

The notion of a neutral public sphere, therefore, is misguided (Fraser, 1992). From its very beginning, the representations or spectacle of neutral publics were only thinly veiled power grabs that functioned to delegitimize the interests of oppressed groups. This grip of exclusivity has, however, changed in Western culture since the time of the French and American revolutions. The vision of publics after the French and American revolutions has been far more porous than its predecessors (Benhabib, 1992), welcoming a greater number of groups into public discourse.

Supporting this trend of increasing plurality is another objective that we observed in our research on the practice of civic designers in the United States. In one case, we learned of civic designers attending to the role that language and style of discourse play in supporting or hindering participation in public issues. At the City of Boston's Housing Innovation Lab, policymakers wanted to explore the viability of compact living models (sub 350 square foot dwellings) for residential development as a way to address the housing shortage. While housing officials and policymakers might be able to grasp what living in 350 square feet means, they recognized that conceptually, it is hard for people who are not architects or builders to respond meaningfully to the question of whether or not they support city policy to pursue compact living as a means of addressing the housing crisis. To bridge this conceptual gap, designers in the Housing Innovation Lab created a model unit that they brought around to various neighborhoods in Boston. Residents were invited to tour the unit and provide feedback about it as a possible direction for the city. In this case, dialogue with government was constructed in such a way that welcomed visceral and emotional reactions, encouraging residents to feel and experience what a potential policy might imply. Designers encouraged visitors to the space to speculate and imagine what life in compact living might mean for themselves and their relatives, using such feedback as valuable insight on how to consider next steps for future policy design around compact living.

That publics have been negatively impacted by this perceived neutrality and binary classification is of particular importance to the civic designer. While the trend of modern publics has been to become more porous and push back against the binary distinction of public and private, civic designers today have a particularly large burden to bear when considering how their work may promote or restrict inclusive behavior. The motivation of design is, in part, one of creating conditions that can bring order out of chaos and help people along a more seamless and less challenging path from point A to point B. But through design one inevitably excludes other possibilities in the favor of a few. This is what design implies. But if design excludes, how do designers contend with this reality in the public setting, and how can they avoid repeating histories of exclusion that have systemized class distinction and oppression? While there is no silver bullet in how to address this, the following sections offer historical precedent for how oppressed groups have contended with this exclusion. Such precedent may offer insight into the civic consciousness designers adopt when doing their work.

COUNTERPUBLICS

Because of the cultural and economic exclusions from participation in public life, Fraser notes that members of marginalized social groups throughout history have created alternate publics, or *counterpublics*—"discursive arenas where members of subordinated social groups invent and circulate counter discourses" (1990, p. 67). Fraser points out that such counterpublics are essential to democracies because they expand discursive spaces, ensuring that the concerns and issues of marginalized groups can come together and be heard.

Participants in counterpublics are the people we have thus far described in previous sections as being excluded from public discourse. Most notably, the distinction based on gender has produced some prominent examples of counterpublics, where women carve out spaces of encounter to share their concerns and find alternative pathways to influence policymakers. As Mary Ryan states, women during the middle of the 19th century in North America who were denied access to the public sphere found "circuitous routes to public influence" (1992, p. 284). One particular issue that rallied women together was the emerging discussion around the regulation of prostitution, which women viewed as legitimizing sex trafficking. Modes of circulating concern about this issue included the creation of female-run charities and women-only voluntary associations designed to meet the distinct public needs of women. Ryan (1992) speaks to the strategy of such associations, describing how they:

> ... held public meetings calculated to influence public opinion on the subject. At other times they directly petitioned the state legislature and appeared at city council meetings, arguing their position. In other circumstances they chose to use the private techniques of lobbying legislature and capitalizing on their personal contacts among public officials. (p. 281)

One such association was the National Woman's Suffrage Association, started by Elizabeth Cady Stanton and Susan B Anthony. In describing the impact of this autonomous public, Ryan (1992) states that "the women's rights movement proceeded to identify a broad agenda of gender issues for black and white women, not just suffrage but also marriage reform, equal pay, sexual freedom, and reproductive rights" (p. 282).

Counterpublics are defined by the nature of the populations they include, but they can also be identified by the modes through which ideas are circulated. Because publics depend on the circulation of ideas through various media, access to the means of producing and circulating media is essential for their formation. In a broadcast context, the high cost to produce and distribute content on the scale of mass media makes it prohibitive for anyone that does not have the financial capital required to participate. For this reason, affluent people have had exclusive control over determining the priorities of media systems, leaving those without the financial means outside the sphere of influence (McChesney, 2004). Furthermore, advertising as a revenue-generating model for television and news requires that shows have a broad appeal so that advertisers can be sure to reach their target demographic (McChesney, 2004). In this mass media context, creating popular content requires avoiding topics that are relevant to smaller subsets of the population. Additionally, content that has a wide appeal often aligns with the interests of media owners and their allies. Citing research by the Glasgow University Media Group, Atton (2002) notes that "An elite of experts and pundits tends to have easier and more substantial access to a platform for their ideas than do dissidents, protesters, minority groups and even 'ordinary people'" (p. 17). As a result of these defining constraints, social issues, interests, and perspectives that do not have a popular following are often left out of mass media content. As such, counterpublics and the modes of circulating ideas outside of these mainstream models take the form of alternative media. This includes everything from pamphlets, zines, pirate radio, and a range of online platforms—any means capable of working free of constraints to ensure the free flow of ideas and perspectives that are missing from mass media.

Focusing on what she describes as the enclosure of media production and distribution created by corporations, Stefania Milan (2016) unpacks the concept of emancipatory communication practices, or distinct technological tactics and strategies for bypassing mass media models of production and distribution. Such practices create "alternative spaces of communication where freedom of expression, participation, and self-organization are practiced independently of social norms and legislation" (p. 109).

Milan points to such examples as Indy Media, which originated during the protests of the 1999 World Trade Organization (WTO) meeting in Seattle to cover the global justice movement. Indy Media, or the Independent Media Center, started in Seattle using an online open publishing platform to share reports and content generated by activists and journalists on the frontline of the WTO protests. Indy Media is now a global network of local centers, with each center acting as a clearinghouse for the news.

Publics come into being when shared interests emerge and circulate in a discursive space with an end goal of influencing policy or decisions. Benhabib (1992) points out that this definition of publics suggests that there are as many publics as there are controversial debates. As a response to the idea of neutrality and bracketing, such a definition of publics may be the only way one can make sense of living in a pluralistic society. And what we have been calling counterpublics, which Fraser (1990) defines as the "parallel discursive arenas where members of subordinated social groups invent and circulate counter discourses to formulate oppositional interpretations of their identities, interests, and needs" (p. 123), might be better classified as simply other publics. The public sphere is composed of multiple publics, each with varying interests and means of communicating them. And while some publics achieve dominance through political and commercial reinforcement, and some are resistant to them, the notion of counterpublics implies a simple dialectic that doesn't exist. It is far more productive to speak of a plurality of publics, each with its own texture and circumstances.

Participation in the public sphere means being able to speak in one's own voice on behalf of one's concerns and those that share them. Making room for a plurality of publics enables subordinated groups to introduce different ideas to the state; and it also offsets the unjust participatory privileges enjoyed by some members of dominant social groups (Fraser, 1990). Doing this is what Benhabib sees as the primary site of struggle in the modern world: the fight to take up issues that are not seen as important in the sphere of public discourse. We see the work of the civic designer as taking up this fight to support the creation of sites where public discourse can exist.

Take, for example, the Invisible Institute's Citizen Police Database Project (CPDP), an initiative based in Chicago designed to surface reports of police abuse. Where this had traditionally not been a matter of public conversation, the public release and hosting of documents regarding police behavior have emboldened those who had been too afraid to speak due to fears of retribution. As Jamie Kalven, the project's founder, described to us in an interview, the project created the space for people to speak up who didn't have it before.

> My own practice is that I've always assumed that my principal allies were not my, you know, politically inactive liberal friends who assume politics is a kind of spectator sport. But a kind of fifth column of folks within the bureaucracies who want to do the right thing, but don't have a lot of space. (personal correspondence, 2017)

The civic designer, by virtue of working with publics, must have an understanding of how and why publics form. Publics operate in and for themselves, but they are often pushing up against other publics in which they have been excluded. To design in the civic space requires not only an understanding of the nature of a problem space, but also an understanding of who defined the problem, for what purpose, and to whose ends. But the sphere of influence is rarely so straightforward as a corrupt politician or a sinister board room. Modern power is diffused, not always dependent on active and visible oppression from the government, but distributed to the governed to govern themselves. Foucault (1995) uses the metaphor of Jeremy Bentham's panopticon, which is a prison design wherein the guard tower is in the middle of the structure, with cells around the perimeter. The guards are looking out of narrow slats in the tower, so that they can see the prisoners, but the prisoners cannot see the guards. Foucault was not particularly interested in prison design, but he saw this particular prison as a metaphor for the function of power throughout modern society. No need for public executions to control the population; the people govern themselves because they never know if they're being watched. This distribution of power is considerably more efficient than models of active suppression, Foucault says, which is precisely why it becomes the dominant mode of discipline in the 18th century. With power distributed, states can simply get more done because there is less need to devote resources to displays of power. People will regulate themselves, not for fear of getting caught per se, but for fear of violating the social order.

NETWORKED PUBLICS

Civic design is concerned with who has a voice and who does not. While progress has been made in much of the world around challenging criteria of participation based on gender and class, there continues to be limitations on who can and cannot be heard. This is determined largely by who owns the means of circulating discourse—in other words, who owns the media. One challenge that many media scholars will be familiar with is that of media consolidation, where in the United States 90% of all content is produced and distributed by only six corporations (The 6 Companies That Own

(Almost) All Media, 2017). The Internet was supposed to usher in an era of democratized communication models (Weinberger, 2015), but companies like Facebook and Google have become part of a small handful of digital platforms that host the bulk of online discussions globally.

Consolidation and control of infrastructures essential to the circulation of public discourse become problematic when the interests of the few who own such infrastructure take precedence over the many who depend on such infrastructure to be heard. At first glance, social media platforms may appear novel in their appearance, but on closer inspection, they are only new manifestations of old power dynamics. Instead of familiar gatekeeper models of media control, where a few networks or movie studios determine what can and cannot be produced, new media economies appear to be open to any content creator. "Filter then publish" is replaced with "publish then filter" (Shirky, 2008), providing the appearance that content creators compete in a free marketplace of attention without the power structure of established gatekeepers. But, of course, things are much more complex than this; social media platforms are not neutral. They are able to control discourse, not toward any ideological ends necessarily, but to the ends of enhanced participation and advertising revenue, through algorithms that present to users precisely what they think they want to see. This is great for markets, but less great for the persistence of a plurality of publics.

The 1990s saw the rise of the Internet and a great deal of excitement around the emancipatory possibilities for freewheeling and unconstrained communication. As long as one had access to an Internet connection and could write, the Internet afforded the tools to publish and share ideas. The lack of gatekeepers was a touchstone of what excited many early intellectuals and scholars of the Internet. The Cluetrain Manifesto (2009), written by one such group of early Internet pioneers and intellectuals, praised the disintermediated nature of communication, unhindered by editorial oversight or the pretense of class. On the Internet, you had to prove your ideas through good argument. An eighth grader could argue with a tenured professor; all that mattered was that she had good ideas and knew how to defend them.

In his account of the early days of the Internet, journalist and Internet scholar David Weinberger (2015) describes his enjoyment of publishing stories on his blog and following the blogs of his peers. They would comment on each other's work and encourage viewers to visit their friends' blogs. As larger news outlets began to catch on to value of the Internet, relationships formed where high-intensity chatter among the network of individual blog publishers would get picked up by a larger news outlet, elevating what may have been a more obscure interest to a more prominent status. This networked relationship between high-visibility news outlets and small-scale publishers created what Yochai Benkler (2006) describes as the networked public sphere, where the lowered costs of digital publishing technology and the potential for pushing content across networks of shared interest created new possibilities for discourse to emerge. Indeed, there are countless stories of the networked public sphere exhibiting moments where critical mass emerged on the level of small blogs and midlevel websites focused on issues and information critical of the state. Once they were picked up by larger news outlets, they held state actors accountable for their actions.

Fast-forwarding to today, this networked relationship between small websites and blogs has waned, giving way to centralized communication platforms such as Facebook, where nearly 1.5 billion people visit daily to share information. While the ease of finding one's peers on a single platform is appealing, the gravitational pull that such platforms have for becoming the preferred tool of communication creates a communication phenomenon that is fraught with difficulties.

As the 2016 elections in the United States demonstrated, platforms like Facebook and YouTube are targets for disinformation, with individuals and foreign governments creating sophisticated campaigns that take advantage of the platform technology to target the preferences of individuals and sway them to engage with content that can be damaging to public discourse.

Prior to 2016, scholars recognized the increasing role that social media platforms played in shaping how people communicate. As Jose Van Dijck describes it, such platforms become an active mediator and shape the contours of public discourse (2013). Tarleton Gillespie (2015) points out:

> Social media platforms don't just guide, distort, and facilitate social activity, they also delete some of it. They don't just link users together; they also suspend them. They don't just circulate our images and posts, they also algorithmically promote some over others. Platforms pick and choose. (p. 1).

The problem of fake news in the 2016 election revealed that these platforms occupy a central role in how content is circulated and discussed. The way that content is distributed and shared plays a significant role in shaping how opinions are formed and subsequently mobilized into political action. And while Facebook has denied its classification as a media company, its decision to allow damaging content to flourish with little oversight demonstrates that they are making editorial decisions that matter quite a bit for the functioning of democracy. According to Mark Zuckerberg, "I consider us to be a technology company because the primary thing that we do is have engineers who write code and build product and services for other people."[1]

And yet, despite continued attempts by Facebook to reject the editorial responsibility of being a media company, they traffic in content and are motivated by profit.

If the infrastructure for publics is put into the hands of private companies that operate on the value of supporting content and discourse that generates profit over public good, civic designers must accommodate this reality into their work. This is not to suggest that they are turning their backs on these platforms, but that they recognize them as value-laden spaces that are not necessarily aligned with the goal of public discourse. And while it is nearly impossible to imagine publics forming outside of these platforms, it is equally as far-fetched to imagine that these platforms provide the infrastructure necessary for a plurality of publics to persist over time. Civic designers are embracing and rejecting the dominant infrastructure of public discourse in a digital culture by actively reflecting on the material realities of the formation of publics.

Engaging with publics is the core competency of the civic designer and the necessary starting point for meaningful inefficiencies. In the next section, we provide an in-depth case study of a design process in New York City, where we worked with a group of delegates in the New York City participatory budgeting process, to help them define the issues and the nature of their assembly. This case study represents a part of the process of creating meaningful inefficiencies, where the designer works in partnership with forming publics to help craft the conditions for discourse.

ROLE-PLAYING GAME
FOR DESIGNING PUBLICS

In 2015, we had the opportunity to partner with the Participatory Budgeting Project (PBP) in New York to incorporate a game we had developed at the Engagement Lab into the delegate process (Gordon et al., 2017). We were invited to do this by Josh Lerner, the author of *Making Democracy Fun* (2015) and the executive director of the organization. Having just written a book about the connection between games and democracy, he was keen on exploring the use of a game in the participatory budgeting (PB) process. In 2013, we developed a role-playing game called @Stake to facilitate dialogue between Romanian and Russian speakers in Moldova to find common ground and seek policy solutions about youth unemployment.[2] We continued to use the game as a tool with diverse and divergent groups as a means of building empathy across stakeholder positions.

Figure 2.1. The @Stake card game being played during a participatory budgeting process in Boston. (Courtesy of Engagement Lab, 2015)

When Eric met Josh Lerner at a conference, @Stake seemed like a perfect tool with which to experiment (see Figure 2.1).

Participatory budgeting is a process whereby local delegates deliberate and have direct say on the allocation of funds. In a typical PB process, a municipality would earmark discretionary funds to be distributed during its regular budget cycle. Instead of budgetary decisions being made exclusively by political representatives, over the course of several weeks, PB delegates deliberate about community needs and priorities and, through a direct voting process, determine how the money gets spent.

The challenge with all civic participation processes, and certainly PB is not immune, is the representativeness of the deliberating public (Barber, 1984) and the ability for participants to understand divergent perspectives. So the value of introducing @Stake at the beginning of the process was to encourage participants to embrace the plurality of voices and interests present.

The game was designed to create a space for emerging publics to experiment with deliberation. Games, as we will describe in the following chapter, can create productive and safe spaces for vulnerability. So, while the stakes are high in a PB process, entering into a space of play removed from everyday life affords people the opportunity to fail with minimal consequences. This reduced pressure to perform can potentially open people up to discovery and allow them to listen to other voices without fearing that it strips them of political capital. This would be true of any game, but the value of @Stake in particular, was that role play and deliberation were the content and core mechanic of the system.

In @Stake, each player gets a character card that has a bio and a secret agenda (see Figure 2.2). For every round of the game, a

Figure 2.2. Range of cards in the @Stake game. (Courtesy of Engagement Lab, 2015)

particular issue frames the conversation and each player has to pitch an idea from the perspective of his or her character while also trying to weave in the character's secret agenda. One player acts as the decision maker who picks the winning idea and awards tokens to that player. Bonus points are awarded to any player who gets his or her secret agenda included in the winning idea. After three rounds, the player with the most tokens wins.

As an introduction to the participatory budgeting process, the game encouraged people to form new relationships and facilitated the generation of new ideas for the actual process, as well as revealed some potential complicating power dynamics. One player described the outcome of the experience playing the game as one of generating more empathy for others in the public forum.

> Because it made me think: okay, how am I going to respond to this in a real-life situation? To me that seems like that's the whole point is you're going to have all kinds of different members of a community in the end, rather than being there advocating for your own thing, use your moment to not only listen to someone else, but to understand from what perspective they come. And then being able to shut up and listen and put yourself in their shoes, and think about what they are advocating for, what their whole life might be like, what their professional life might be like, why they are here advocating this. I think a legitimately done, polished response, is the most valuable aspect of that game. (personal communication, March 2, 2015)

Another player noted the effect of the game in orienting him toward a more empathetic perspective while also preparing him to best approach the task of researching and advocating for his ideas.

> I thought it was a good preparation for making a case for a project. This is what we ended up doing at the budget delegate meetings: we did a lot of research, talked about the project and described it, made a case for it. It was good preparation.

I think people represented different interests. Where people were coming from was interesting, and it helped us understand people's interests. (personal communication, February 4, 2015)

While many people were receptive to the idea of trying out the game, others where not. One group had organized ahead of the meeting and, as we learned, were planning to push their ideas through the process. Upon learning that they would participate in a game ahead of the process that encourages dialogue and empathy, they left the meeting. Although it is not known if it was the cultivation and empowering of participants with empathy, we see how Habermas's concern about how entrenched interests can disrupt the potential of discourse to achieve democratic parity plays out in this example. While the act of organizing people to represent a shared interest is not bad, the intent of doing so to crowd out dissenting or varying viewpoints is. In such an example, we see how presenting moments where people are oriented toward a model of discourse that gives primacy to empathy and listening, even if it takes more time to accomplish, can present itself as a threat to models of public process where efficiency becomes aligned with lopsided power dynamics (see Chapter 4 for a more thorough analysis of this moment).

The work of preparing people, from the equally important perspectives of process and building empathy, helps to mitigate an outcome where the participatory budgeting process doesn't live up to its fullest potential. As we describe elsewhere, "When people are asked to move immediately into deliberation with real stakes, even when the process is heavily structured, lacking the empathetic common starting point, hinders one's ability to generate new ideas" (Gordon et al., 2017, p. 3799).

We use @Stake as an example not to celebrate the function or value of games in civic process (Lerner, 2014), but rather to allow the game to display the work involved in the formation of publics. A decade prior to this work, we wrote about the virtual environment *Second Life*, specifically regarding its capacity to tease out lifeworlds, as we experimented with its use in a community planning process (Gordon & Koo, 2008). The importance of these tools for the work of civic design is the active construction of discourse spaces for the formation of publics and, specifically, ones that embrace the media platforms that influence them. As we have been describing in this chapter, the means of communication that groups of people use to connect is impactful on the resulting shape of the public that emerges. The appetite for digital tools in public participation processes is growing, largely because the demand for greater scale and enhanced efficiency is growing. So even though @Stake was designed as a conversation tool, it is still often looked at as a surveillance tool. As with any civic tech used in public process, people are eager to "see the data." The promise of digital tools for many planners is the archivability of conversations, not the emergence of discourse. It is for this reason that we deliberately avoided collecting any "game data" and instead allowed the tool to be generative. Even the digital version of the game avoids the collection of conversation as data, as publics are rarely, if ever, the sum of their parts.

PUTTING PUBLICS INTO PRACTICE

A meaningful inefficiency is an approach to civic design that seeks to transform institutional logics through the structure of publics, play, and care. What we have described in this chapter is the structure and nuance of publics, the subject and raw material of every civic design process. We have provided an example of a tool that was brought into a process at the beginning in order to make explicit the work of forming and communicating publics. This can happen in many different ways. The point of sharing this example is not as a best practice, but rather as a demonstration of the value of doing the work in the first place. Too often in civic life, the existence of publics is perceived to be a given, emerging as already formed into a healthy public debate. But in fact, the way in which they form, the struggles they embody, are an important consideration for the designer, as ignoring these parts can result in damaging normative assumptions about who and what comprise publics.

We spoke to dozens of practitioners doing this work in a variety of ways, primarily undertaken through the four activities: network building, holding space for discussion, distributing ownership, and persistent input. So while this chapter has focused only on the formation of publics, these activities are distributed throughout the design process. Andrea Hart of City Bureau, a community news organization in Chicago, describes her work in creating a participatory news platform:

> What does it mean to also empower folks to have ownership and contribute to information systems? We talk a lot about changing systems of journalism so that we hear folks. Journalism shouldn't be giving voice to the voiceless. It should be thinking about what are the methods in which we are collecting voices, and how are some of those methods deaf or just not listening well? (personal communication, 2017)

She is talking about both holding space for discussion and distributing ownership to participants, each in service to the formation of publics. She understands quite well that local news is a platform, in need of design, that is not an end in itself, but a mechanism that creates the conditions for publics to form. @Stake, while a temporary utility to aid in the formation of publics, had similar aspirations to "listen well." The story of the game's use in the PB process demonstrates an attempt for inchoate publics to listen to each other. It had little to do with the designer listening to an already formed public, and everything to do with allowing participants to hold space for each other.

Perspectives can be excluded from public discourse in a number of ways. Something as simple and straightforward as the time or place in which public hearings are held can have implications on who participates in dialogue around issues of public concern. In their exploration of new approaches to contend with the housing shortage in Boston, designers from the Housing Innovation Lab recognized that, if they followed the standard model of holding community meetings, they would miss wide swaths of resident perspectives.

> The conventional model is a community meeting, which might be at 6:00 on a gym on a Wednesday night, and you might get a certain number of people out. Typically you get a kind of a—a typical goer—somebody that goes consistently. You might not get a lot of people, you might not get somebody that is—cares about their community, but doesn't necessarily have the time to go. And, so, you miss out on a lot of people. And, so, part of this was about getting—accessing those people that aren't necessarily at the community meetings. (Max Stearns, personal communication, 2017)

The response to this constraint was to bring the conversations via the simulated Urban Housing Unit model (described earlier in the chapter) to different neighborhoods around the city, locating it in places and times where they knew they could intercept people during their daily routines rather than forcing them to go out of the way. Locations such as major transportation hubs or public events were used so that people could encounter the installation in a more seamless fashion. Once inside the unit, residents were able to respond to questions researchers had about the topic of compact living as well as explore facets of the topic that were most relevant and connected to their lived experience as residents of Boston.

The technique of offering a common object of curiosity to forge a public is not uncommon. The Nomadic Civic Sculpture, by the artist Salvador Jimenez-Flores at the Urbano Project, was an installation that moved from one public park to the next in various Boston neighborhoods, featuring a surface on which people could write responses to questions identified as relevant to that neighborhood.

The questions and opportunities for interaction on the sculpture evolved and were refined as Jimenz-Flores and the young people working at the Urbano project responded to the answers written on the surfaces of the sculpture. In this way, they created an object designed to draw out latent concerns, using existing knowledge of the neighborhood but not predetermining the framework of the conversation, using instead prompts and probes that encourage people to shape the details of the dialogue. Subsequent events with the project included holding public screen-printing workshops as a way to produce new media, from t-shirts to posters, that helped broadcast and raise awareness around the issues which the sculpture's installation in various neighborhoods helped to articulate.

As Jimenz-Flores describes it, bringing the installation and the opportunity to engage in a conversation about civic issues to a community in locations that are convenient was an essential part of the project's success (see Figure 2.3).

Figure 2.3. *The Nomadic Civic Sculpture* by artist Salvador Jimenez-Flores at the Urbano Project. (Courtesy of Culture House)

There is a real power in saying, we're going to go to the grocery store, and we're going to be in the parking lot of the grocery store because we know that's where you go. So I think that was one of the main goals, to be able to bring the work to the people instead of asking people to go and come to us to see the work. (Salvador Jiminez-Flores, personal communication, 2017)

A focus on location and demographics can be an antidote to an unintentional bias produced by the dynamics of social networks. For instance, in their initial work to recruit a network of residents in Chicago to test out and provide input on civic engagement software, the Civic User Testing (CUT) Group relied on the contacts of their executive director who had strong relationships with people in information technology and government living in the North Side of Chicago, which was predominantly White and working class. As Kyla Williams, current interim executive director, describes it:

When we did the initial cry out for participants, of course the areas that filled up were the ones that were segregated based on demographics. We're on the grid system. North, South, and West is where the folks live, and on the North side typically is more white middle and working class folks. South and West is normally black or brown, typically lower social economic conditions. (Kyla Williams, personal communication, 2016)

While this recruitment yielded an important section of the Chicago population, it also left out two thirds of the demographic makeup of Chicago. It left out people who did not work in the technology and public sector. To address the gap in perspectives, Sonja Marziano, the project coordinator at CUT Group, and her team targeted populations in neighborhoods that were missing from the initial response so as to create a more representative sample of the population of Chicago.

We mapped out to all of the libraries, community service centers, hair salons, barber shops, schools. Places and faces where we knew regular people would be, and mapped out based on those central points. Because a lot of those central points either had a shopping district or had an entertainment venue or something that was at least indicative to us that they had a lot of people, potentially, that could go back to other places. So, we mapped out this strategy based on the map itself, and we decided that we were going to go very low tech. We took fliers and sign-up sheets out in the community with a person who basically got a street team—very similar to what one would do when they're flyering for a political party. We took a street team of individuals who flooded the black and brown communities and some of those northern communities where there were different languages spoken—Polish, Spanish, and Mandarin—and basically just walked the streets using our back-ended map as a way to try to reach more people. (Sonja Marziano, personal communication, 2017)

Like the work of Andrea Hart at City Bureau, Sonja and Kyla at CUT Group recognize that achieving inclusivity in public dialogue requires the very deliberate work of first identifying who needs to be at the table and then finding ways to make sure there are as few hurdles as possible to getting them there. Similar to the strategy of deploying the Nomadic Civic Sculpture, this means going to the neighborhoods of target demographics and engaging people through means in which people discover opportunities to participate that fit into their everyday routines.

These examples shed light on how designers are working with publics to establish boundaries and identities, even prior to doing the work they formed to do. Finding where people are, assuring that people feel invited into the conversation, and creating the conditions for negotiation is the hard work of civic design.

In the next chapter, we examine the role of play in civic design. We look to the function of play in everyday life and argue that play functions as the precursor to action taking. While we talk about games as systems that structure play, we do not position them as privileged or in any way ideal systems. They give us structure to have the conversation about the role of play in civic design, and they allow us to talk in concrete terms about effective strategies of designing meaningful inefficiencies.

[3]

PLAY

Every generation, the indigenous peoples of the Shuswap region of British Columbia intentionally move their village. As the Canadian ethnographer Richard Kool described to psychologist Mihaly Csikszentmihalyi, the region is considered:

> . . . a rich place: rich in salmon and game, rich in below-ground food resources such as tubers and roots—a plentiful land. In this region, the people would live in permanent village sites and exploit the environs for needed resources. They had elaborate technologies for very effectively using the resources of the environment, and perceived their lives as being good and rich. Yet, the elders said, at times the world became too predictable and the challenge began to go out of life. Without challenge, life had no meaning. So the elders, in their wisdom, would decide that the entire village should move, those moves occurring every 25 to 30 years. The entire population would move to a different part of the Shuswap land there, they found challenge. There were new streams to figure out, new game trails to learn, new areas where the balsamroot would be plentiful. Now life would regain its meaning and be worth living. Everyone would feel rejuvenated and healthy. Incidentally, it also allowed exploited resources in one area to recover after years of harvesting . . . (Csikszentmihalyi & Csikszentmihalyi, 1992, p. 184)

This is a meaningful inefficiency—the intentional cultivation of difficulty, confined within clear structure, for the purpose of meaning

Meaningful Inefficiencies. Eric Gordon and Gabriel Mugar, Oxford University Press (2020). © Oxford University Press.
DOI: 10.1093/oso/9780190870140.001.0001

making. Even though there were clear environmental reasons, namely the recovery of resources, for moving about the region, the elders told the story of renewing challenge so that life would be worth living. So, what would appear to be a hardship is in fact a kind of playfulness. Challenges are cultivated for the sake of challenge, with an understanding that the process of overcoming those challenges, not the result, is meaningful. John Dewey writes about the affordance of play in a democratic context: "Persons who play are not just doing something . . . they are *trying* to do or effect something, an attitude that involved anticipatory forecasts which stimulate their present responses" (1922, 238). Play, for Dewey, is directional and motivated. Whether the goal is to win a game, exist within an imaginary world, or simply to engage in an activity for its own sake (i.e., tossing a ball), play is presence structured by futures or other time-bound limitations. Public life is filled with future orientations—traveling to work, accessing services, even voting—but it typically is absent of presence, activities contained by futures, but which exist for their own sake.

In this chapter, we present play, or those cordoned off spaces that exist for their own sake, as a necessary component in the design of meaningful inefficiencies. We have already established that civic design must cultivate and involve publics. We now describe the benefits of designing processes through which publics can play, so that they may take action that results in care. We begin by reviewing the philosophical treatment of play and examine where it directly applies to civic design. We then explore a short case study of a project that sought to involve youth from underserved neighborhoods of Boston in the sourcing and writing of Pokéstops in the augmented reality game *Pokémon Go*. This project was an example of a private–public partnership designed as a meaningful inefficiency. We will discuss the difficulty in aligning goals across stakeholders and some of the challenges such design processes pose. Finally, we explore how public-sector and civil society organizations have put play into practice through gamification efforts, and we address some of the challenges with these approaches. We look closely at why people choose to use games and how some of the assumptions about the work games do provide a disconnect from the value of play.

PLAY AND LIMITATIONS

The integration of play into the structure of society is, in most aspects of modern life, a rarefied occurrence. According to animal play theorist Robert Fagan, "play taunts us with its inaccessibility. We feel that something is behind it all, but we do not know, or have forgotten how to see it" (quoted in Sutton-Smith, 1997, p. 2). Certainly, there are prominent places for play in modern society, namely child's play, or sport for adults. Except for competitive sports that masquerade as a form of productive labor, play is set aside from "serious" matters. It is hidden in plain sight, and apparently only children know how to see it. And what's worse, when it emerges from the protective padding of child's play, it is often characterized as an escape from, not an organization of, the structures of modern life. "All serious activities," writes Hannah Arendt in *The Human Condition*, "irrespective of their fruits, are called labor, and every activity which is not necessary either for the life of the individual or the life process of society is subsumed under playfulness" (1998, p. 127). According to Arendt, playfulness is relegated to the merely excessive and the frivolous.

Arendt's critique of modern society is that play, as opposed to being a kind of action (described in the Introduction), functions instead as a safety valve for a modern laboring society, where alienation from work or the outcomes of one's labor are normalized, and the *vita activa* is reduced to a simple binary between labor and play. In an extended footnote in *The Human Condition*, Arendt equates this to another binary between necessity and freedom, wherein everything becomes either an activity to meet one's needs (i.e., we work in order to be able to live) or to be free from that necessity. Play enables what Arendt calls the laboring society, where purpose is relegated to the paycheck and the accumulation of capital. Not even the "work" of the artist is left, she says: "It is dissolved into play and has lost its worldly meaning" (1998, p. 128).

Worldly meaning, and by extension "the world," comprises all the ways in which people interact with each other and community. What Arendt famously called "dark times" describes a state in which people are increasingly alienated from the world, or the structures that comprise the space in between individuals. Dark times are not void of play. In fact, one could argue that play makes dark times palatable, even acceptable. Playing can connect people to the world through action; it can generate discourse not about individual pleasure, but about the "world," or the structures that comprise the space in between individuals. As a means of reclaiming the *vita activa* and its relation to the world, play can be imagined as part of action taking, removed from its function as a mere retreat from the world, and brought into the process of *making the world*.

Standard narratives of innovation are focused on "earth-shattering" discoveries and novel interventions (see Chapter 1). They are world disrupting, not world making. But civic innovation is different. It implies the structuring of a world that enables a multitude of publics to take action. It is not just about problem solving, but about cultivating the conditions in which problems can be solved. This space for play can be any clearly demarcated space (literal or metaphorical) whereby people have permission to play as defined by the imposed or self-assigned boundaries. Innovation, in the civic sense, is not defined by the crafty actions of an individual player in a complex game, but rather by the design of the game itself.

The sociologist Pierre Bourdieu does not exactly address the quality of playing, but he uses the metaphor of a game to describe how people interface with social systems in general. He begins by referencing the common phrase of having a "feel for the game" and he explains it as the "almost miraculous encounter between *habitus* and a field" (1990, p. 66). The field in this case is a playing field, or a board, the "magic circle" of a game—that discrete space that a player steps into in order to play (Huizinga, 1955). And the *habitus* describes being in the world—all the normative assumptions and ingrained behaviors that define individuals and cultures. What Bourdieu is describing is that experience when naturalized practice butts up against systems. When one has a "feel for the game," it suggests that the game system has become natural for the player, or in the practice of playing the game, she no longer feels its artifice. Extending this metaphor, he states that in social fields, "which are the products of a long, slow process of automatization, and are therefore, so to speak, games 'in themselves' and not 'for themselves,' one does not embark on the game by a conscious act, one is born into the game, with the game" (1990, p. 67).

He continues, "the earlier a player enters the game and the less he is aware of the associated learning (. . .), the greater is his ignorance of all that is tacitly granted through his investment in the field and his interest in its very existence and perpetuation and in everything that is played for in it" (1990, p. 67). As a good sociologist, Bourdieu is not talking about games; he is talking about society. The game is a metaphor through which to understand the tacit acceptance of rules that govern daily life, and that lead to ideological complacency, and even oppression. Being in the game can be blinding. When the "feel for it" becomes too natural, then the player fails to see the artifice, or the structures that surround and support the process of play. By bringing play into the civic design process, the goal is to counteract the moments when the visibility of power structures fades into the background, operating as a means to bring the questions of process and power in as key variables influencing the design of civic spaces.

The rules of the game are policed both by the players and by those looking at the game from the outside. In professional baseball, for example, the players are on the field and fully committed to playing, but so too are the spectators committed to their play. They enforce the rules by debating who's the best, pouring over statistics, and telling stories, all within the carefully limited context of the game. The anthropologist Norton Long describes how this process creates the conditions through which sense can be made of the situation.

> If we know the game being played is baseball and that X is a third baseman, by knowing his position and the game being played we can tell more about X's activities on the field than we could if we examined X as a psychologist or a psychiatrist. If such were not the case, X would belong in the mental ward rather than in a ballpark. The behavior of X is not some disembodied rationality but, rather, behavior within an organized group activity that has goals, norms, strategies, and roles that give the very field and ground for rationality. Baseball structures the situation. (quoted in Tsebelis, 1991, p. 33)

Games provide context for understanding actions that individuals take within a given social context. Each action taken by the third baseman is structured by the formal rules of baseball (i.e., play will take place over nine innings, each inning is comprised of three outs, and each batter is limited to three strikes before getting an out); it is also structured by informal norms (the third baseman typically stands between third and second base, and if the ball is hit to him with no outs, he would typically throw across the field to first base). When one is observing a game, one is aware of some combination of formal rules and informal norms, but in all cases behavior is structured within limitations. So while games represent systems that constrain, they also represent systems that enable acceptable play. They provide opportunities, within a prescribed set of rules and norms, to play toward desirable outcomes. This goes back to John Dewey's forecasting, or the ability of play to be at once present and future oriented.

This structural approach to play is not without its dangers. Bourdieu warns of the game becoming normalized, wherein one doesn't see the game as fabrication, but as naturalization. What if the baseball player didn't know that the rules of baseball were limited to the field? Would he be playing? Or simply allowing the rules to play out before him? Play, in this case, is different from the game. If the game does not have clear limits, then it becomes impossible to play within it. Games provide structure for play. But a game can cease to be a game when the structure fades either through disuse or active resistance. When a group of children are playing a game of "cops and robbers" in the playground and one child who is playing the robber says, "I am now a dolphin," and another says, "I am now a whale," the game rules change and play is redefined. While play is structured through limits, players can modify those limits or simply step out of them. The philosopher Bernard Suits (2005), as quoted in the Introduction, puts it this way: "To play a game is to attempt to achieve a specific state of affairs, using only means permitted by rules, where the rules prohibit the use of more efficient in favor of less efficient means, and where rules are accepted just because they make possible such an activity . . . playing a game is the *voluntary* attempt to overcome unnecessary obstacles (italics added)" (p. 10).

Playing within a game is always voluntary. One must make the decision to step into the magic circle and accept the rules that structure play. However, for Bourdieu, when one develops a feel for the game, and is even perhaps born into it, the rules are accepted and not elected. In this case, the game is not a place for play, but a place where rules play out. The ability to understand that one is playing a game, is acting within a set of rules that are distinct from those governing everyday life, and importantly, that those rules are prohibiting the more efficient in favor of the less efficient means of achieving a state of affairs, are the system requirements for play. Meaningful inefficiencies are fields wherein one voluntarily steps so that she can play, and where the outcomes of such play are apparent but not primary.

The goal of meaningfully inefficient systems is for its occupants to make power and process in a particular context visible by confronting challenges that have become normalized, or invisible.

When playing, one is both inside and outside. The poet Charles Baudelaire, writing of his experience on the Parisian sidewalk in the 19th century, described himself as a *flaneur*, both a part of and alienated from the crowd. As he watches the crowded sidewalk from inside a cafe, he is a spectator of urban play. But when he leaves the cafe, he is again a part of the crowd. While he doesn't use the term, the urban *flaneur* is playing, stepping into play (and its structuring limitations), but always capable of stepping out of it. *In play there is agency.* There is agency in the moment (Huizinga says that "all play means something" [1955]), and in the player's ability to step outside. If the player were to feel manipulated, as if she was "being played," she can step out of the game, refuse to play, or demand a check on the rules. As a design strategy, meaningful inefficiencies structure play, but equally as important, they structure the ability for players (and spectators) to reflect on the rules and limits of a game. By defining the field and its constituent rules, one inevitably defines the conditions that create the experience of playing within it, and the possibilities of playing with it.

Consider the example of the Shuswap people again. By designating another field, defined by its rules and unnecessary obstacles, they are able to play toward achieving a certain state of affairs. Media theorist Ian Bogost further refines the conditions of play. "Play," he argues, "is the act of manipulating something that doesn't dictate all of its capacities in advance, but that *limits* its capacities through focus and exclusion" (2016, pp. 92–93). In his book *Play Anything*, he uses the example of walking with his daughter in a mall while holding her hand. She is bored and dragging a bit, but then she starts playing a game where she tries to walk only on the cracks between the tiles. By self imposing obstacles on herself, she regains agency in the situation through her play. Play is a mechanism of overcoming structural limitations to achieve a goal. In this case, her goal changed from "I want to be home" to "I want to not step off the cracks." One might argue that this is an example of oppression, where the oppressed gives up on real structural change and cultivates false agency. But, while she remains captive by her father's walk through the mall, she now has agency within her self-imposed rule set to explore, experiment, fail, and discover. In fact, her play is comprised of actions—what Arendt would call "new beginnings"—that set other events into motion.

The play theorist Miguel Sicart (2014) calls this "playfulness," which is when people play in a context that is not "created or intended for play" (p. 27). It is about the "appropriation of what should not be play" into something that takes on personal meaning for the player. Bogost's daughter is exhibiting playfulness in her appropriation of nonplay spaces, but in so doing she is crafting new spaces for play that are bounded, confined by rules, and self-imposed.

Civic design always makes room for play and accommodates playfulness. Civic spaces are play spaces. Instead of asking "How does one get from here to there?" the civic designer asks "How does one understand the practice of playing within limitations to achieve goals?" The end result is a system that can be played, or a system that presents opportunities to be shaped by the people within it—where they are allowed to *play within and with* the system itself, not just *play out* preconceived tasks.

PLAY AND FREEDOM

Play is an outcome of particular systems designed with the appropriate structure and room to cultivate it. Bogost explains that to play is "to take something—anything—on its own terms, to treat it as if its existence were reasonable. The power of games lies not in their capacity to deliver rewards or enjoyment, but in the structured constraint of their design, which opens abundant possible spaces for play" (2016, p. x). But play is not just the result of system design, it requires what Suits (2005) calls a ludic, or playful, attitude.

Playing requires a suspension of disbelief and a voluntary stepping into the magic circle. Even in Bogost's example of his daughter being dragged through the mall, she was forced into the situation, but her decision to play was voluntary. Playing a game of baseball requires an acceptance of the rules and willingness to play within them; playing a fantasy game of princesses and dragons requires a playful attitude that both accepts limitations and desires to stretch their limits. Even driving home in traffic can be playful. When Eric (one of the authors) worked in Los Angeles in the 1990s, he lived on the east side and worked on the west side. Each day on the way to work, he would take a different route to beat his time to and from work. The activity was the same, but by transposing rules and obstacles (time) onto the situation, he was able to have some agency in an otherwise dismal situation.[1] Even while physically occupying the same space, the game was a distinct field, in that he would determine prior to departing home or work that he was playing the game, accept that it was reasonable, and then operate within the rules. There was nothing at stake except the satisfaction of beating a previous time, but by providing permission to play, the field becomes a space of manipulation and discovery—of new paths in the environment and a constant reminder of the limitations of the game itself.

It is possible that the traffic game was simply an escape from reality, a binary between labor and playfulness (Arendt, 1998). It passed the time. It was fun. It was amusing and entertaining. But it was also frustrating and enervating, as it focused attention on the severity of the traffic and its subtle variations from day to day. Play, according to Miguel Sicart (2014), is not simply about fun—he prefers the term *pleasurable*. "The pleasures it creates are not always submissive to enjoyment, happiness, or positive traits," he says. "Play can be pleasurable when it hurts, offends, challenges and teases us, and even when we are not playing. Let's not talk about play as fun but as pleasurable, opening us to immense variations of pleasure in this world" (p. 3). Pleasure provides a way of talking about play that is open to a range of human responses—whereas fun too easily degrades into frivolity, pleasure resonates with a wide variety of meaningful activities.

But who gets to play? Is the playful attitude a luxury, only available to those with existing mobility and freedom of choice? By extension, can one talk of pleasure in design when the real challenge is meeting people's basic needs and human rights? This is precisely why, we argue, play is important. Beyond the provision of services, the core requirement of good civic design is empowering people to take action within systems and holding institutions accountable for the rules they create. This is particularly important when the design process involves those typically excluded from systems. McDowell and Chinchilla (2016) advocate for a frame of civic inclusion, which "requires that all individuals learn to engage with established organizational structures, and that institutions become adept in serving an increasingly heterogeneous membership" (p. 462). This, of course, presents a design challenge, which can be met by recommending that institutions "design from the margins." In other words, inclusion only happens when design processes (not just outcomes) include those who have been excluded. In this case, the design process, rather than providing an efficient solution, becomes a tool to motivate the attitudes and knowledge required for inclusion. Playfulness brings freedom to a system through choice—even when that system is oppressive—consider traffic in Los Angeles or being led through a mall by your father. bell hooks defines *oppression* simply as "an absence of choice" and *discrimination* and *exploitation* as being presented with fewer choices than those in dominating groups. Play can respond to oppression by offering choices where before there was none.

Bringing freedom through choice was at the heart of the Place/ Setting project, a series of social art installations that one of the authors (Gabriel) collaborated on with the design firm French 2D. As academics and practitioners, the organizers found that the humanity of idea exchange was often lost to the formality of contexts like symposia, lectures, and conferences. Power dynamics and politics stifle conversations, and the physical, social, and financial barriers to participation prevented access to a broader public that might want to hear the ideas. To address these challenges, Place/ Setting experimented with the physical arrangements, atmosphere,

and formats of spaces for conversations as people explore important social issues. In one version of Place/Setting, the organizers
ran a workshop with students who were a week into a new program. The organizers asked the students to design dueling dinner
parties. They were given tables and chairs and some conversation
starters. How they chose to deploy these resources was up to them.
One group tested out a triangular setup for the tables, while another group created a cozy campfire setting, using the tables as
tents while they sat around in a circle. After the groups finished
their dinner parties, they shared their approaches and results, reflecting on how playful experimentation with the physical and social structure of the conversation created varying levels of human
connection (see Figure 3.1).

There is always ambiguity in choice. Simone de Beauvoir argues
that personal freedom emerges through ambiguity. Building off
of Sartre's existentialism, wherein personal freedom rests in the
movement from facticity, or that which is, to transcendence, or
that which is not yet, de Beauvoir argues for an ethics of ambiguity, wherein one's freedom is not about the perpetual striving
for transcendence, but about the recognition of the ambiguity
between facticity and transcendence. De Beauvoir (2012) argues
that "to attain his truth, man must not attempt to dispel the ambiguity of his being but, on the contrary, accept the task of realizing
it" (p. 12).

Ambiguity does not equate to lack of clarity. De Beauvoir argues
that an ethics of ambiguity implies the acceptance of the task of realizing the ambiguity of one's being. So, not the pursuit of transcendence or the acceptance of facticity, but the acceptance that one will
always strive to be comfortable in between these constructs. For
example, an average waiter in Los Angeles is not merely the waiter
(facticity), nor is he the actor (transcendence), but for him to be
free, he must accept that he is always both. There is clarity in that
ambiguity. Civic life often demands the embracing of ambiguity—
as one might be neighbor, customer, renter, advocate, and visitor

Figure 3.1. Dueling dinner parties in a Place/Setting event.

all at the same time. As opposed to pushing people into one primary role, it is the challenge of the civic designer to understand that
people are multiple. Embracing of peoples' multiple identities and
needs is a step in the direction of designing public contexts that

embrace pluralism rather than operating on a rigid system of inclusion and process that, by design, excludes wide swaths of voices and concerns.

Perhaps the most important contribution of de Beauvoir to this discussion of play is her assertion that the pursuit of freedom is never individual. An individual's freedom necessarily impacts the freedom of others. For example, if backpacking around Europe is one person's realization of transcendence, all the people that retreat into their facticity to support her journey are impacted. The ticket collector on the train, the people whose livelihoods are dependent on the tourist-oriented economies that support the traveler's comfort— these people are in some way impacted by the traveler's pursuit of her freedom. "Only the freedom of others," de Beauvoir says, "keeps each one of us from hardening in the absurdity of facticity" (2011, p. 77). For de Beauvoir, freedom is social; it is in and of the world, not a metaphysical pursuit one undergoes on her own. This is a radical, and feminist, concept that places the pursuit of freedom not only within the constraints of a system that enables individual choice, but within a system where individual choice always impacts the choices of others. While playing a game of baseball, the third baseman is free to play within the limits of the game. But if the second baseman was not free to do the same, then the third baseman would not be free. Simply put, one cannot be free to play a game that others are not free to play. This does not mean they need access to that game in the moment, but that the rules are transparent and accessible so that they are understood both by players and observers. Even a game of solitaire, while not open to other players in the moment, is understood by others as a game with rules that can be played. So, while one may not be able to play in the World Series, one is able to play a game of baseball with friends. In short, for one to be free while playing, the structures of play need to be accessible, either in the moment or at a later time, to anyone who desires to play. The work of civic designers is to ensure that the context they design for is informed by an awareness of such rules and subsequently reveals and makes visible such rules to future participants.

GAMIFICATION

The summation of what we have been arguing in this chapter is that play is the precondition for action taking. And if civic life is the ability for people to take actions, then creating the structures for play is fundamental to civic design. And yet what wins the big grants and press coverage are the tools, the machines that accelerate people's achieving of goals, and not the conditions under which goals are achieved. This is well encapsulated through the example of gamification, defined as adding game elements to non-game contexts (Deterding et al., 2011), or more specifically, the transformation of "non-recreational, tedious tasks, which are often driven by utilitarian motives, into enjoyable, self-purposeful and hence hedonistic activities" (Thiel et. al., 2016, p. 35). From workplace compliance (Nelson, 2012) to physical activity (Zuckerman & Gal-Az, 2014), the use of incentives such as points and badges offers a challenge to make seemingly boring activities meaningful.

Even though the term *gamification* has only recently become popular, there is a long history to game-based incentives (Hamari & Koivisto, 2015; van der Heijden, 2004). Some of the earliest examples come out of the Soviet Union. Soviet work games were designed to "reconceptualize" work away from capitalist assumptions and provide a way of filling the void left by the absent game of capitalist markets. While direct individual competition was not desirable in the workplace, factories built systems to compete with each other, or groups within larger teams were given incentives such as badges and rankings to motivate production (Nelson, 2012). On the contrary, by the 1980s, trends in management consulting in the United States turned to making the individual worker happy by making work fun. Emphasis was placed on individual productivity and satisfaction of the worker. What Nelson calls "funsultants" were brought in to engage the employee with "elements that attempt to make workers feel like they have a stake and expressive role in their workplace" (2012, p. 2). While the difference between socialist and capitalist workplace games was subtle, with the former focusing on collective outputs and the latter on individual agency, each directed play to the service of labor. In the gamified workplace, in both the socialist and capitalist contexts, play was offered as an escape from labor.

The term *gamification* began as a description of advergames in the late 1990s, or flash-based web games that contained sponsor messages. These banner ads were simple games that attracted user attention to an advertising message. They were by no means quality games, but they used the structure of games to capture people's scarce attention. Since these early days, gamification has been widely celebrated as a motivation system in business, health care, and government (McGonigal, 2011). Through the addition of game elements, including points, badges, leaderboards, mayorships, and rewards, gamified systems have been applied to a range of tasks, from worker productivity, to diabetics taking their insulin, to exercise (i.e., Fitbit). As Walz and Deterding (2014) point out in the preface to their edited volume *The Gameful World*, there is a robust debate as to the value of gamified systems. Jane McGonigal captured a popular audience with her TED talk and book *Reality Is Broken* (2011), where she argued that games can repair this broken reality through maximizing individual potential. On the contrary, many critics see gamified systems as taking "the thing that is least essential to games and representing it as the core of the experience" (Robertson, 2010). In other words, gamified systems tend to place extrinsic reward systems onto actions, such that the action is not done for its own sake, but only for the sake of the reward (i.e., points, badges, etc.). This can be quite effective with certain behaviors (Fitbit keeps track of personal exercise and rewards users for "healthy" behaviors; or Duolingo, the language software, provides points and incentives for continuing to use the system). In the case of exercise or language learning, the motivation of individual pursuits has little bearing on the freedom or pursuits of others. And it should be pointed out that, while game elements are used in these systems, it is difficult to talk about them as playful. One does not play the Fitbit or Duolingo; they clarify incentives so users have less room to play. They are designed to be efficient, not inefficient, and they are goal oriented, not concerned with embracing ambiguity.

While gamified systems make good sense for certain contexts, when they enter into the civic realm, wherein play is the precondition for action taking, they can be destructive. Gamified systems are transactional and predefined, preventing players from inventing new approaches to play or interacting with each other in any direct way. As a general rule, they are not playful. As an extreme case, these systems resonate with the cultural critic Herbert Marcuse's (1991) critique of mass culture. He warns against cultural systems that are set up to contain unpredictable human passions—such as love, fighting, awe, wonder—in order to maintain control over people. He calls this function of advanced capitalism "repressive desublimation." By sanctioning a controllable amount of "uncontrol," or allowing citizens to *feel* as if they are engaging in something free and real, they think that they are cathartically releasing parts of their inner natures—that is, desublimating—but they are in fact just incorporating into their psyches the repressive designs of the entity allowing this desublimation to happen. When powerful institutions—from government to corporations—create "fun" processes for participation, they are ostensibly tokenizing or placating their citizens or users. They are looking for them to operate within distinct modes of behavior that meet the needs of the game designer. There are many potential examples of this, from *Commons* (2011), a location-based game that seeks to make reporting more gameful, to *Community PlanIt* (2011–2015), a planning game where big ideas are shared to produce influence toward short-term goals (Gordon & Baldwin-Philippi, 2014). While neither of these systems is designed to placate, often as these tools are implemented, they become means of data collection or placation. Again, as one of the authors of this book was a creator of the *Community PlanIt* platform, we understand the value of the system in certain contexts, but we are also keenly aware of how implementation can remove the nuance required for the successful incorporation of play in civic contexts. Game scholar Jesper Juul makes this point as it pertains to corporate structures. During the 2008 financial crisis, large banks and financial institutions made their organizations "too game-like by giving employees the clear goal of approving as many loans as possible and punishing naysayers with termination. This was a case where the design that works so well inside games can be disastrous outside games. . . . Games are not a pixie dust of motivation to be sprinkled on any subject" (2013, p. 10).

Our contribution to this debate is not to argue for or against gamified systems, or to lament, as many game scholars have (Bogost, 2014; O'Donnell, 2014), the lack of quality games in applied contexts. What we have seen in our interviews with practitioners who have applied games in practice is that they really are interested in designing for play, and the game becomes a shorthand for that process. From a project manager at the San Francisco Public Utilities Commission, who was part of a team that created a game for engaging people in conversation about the local watershed: "We see that people who've gone through this gaming process have gone through an extreme experience of education of how a watershed works, how the sewer system works, and they've done it by learning not just from us, but from each other" (anonymous, personal communication, 2014). The focus on peer learning is important to understand how the priorities of the project allowed the definition of the game to emerge. When the project began, they were calling it a dialogue toolkit and then a charette. "It de facto became a game because people played it," according to the project manager. "And so the difference between just an exercise and a game is this notion of winning—or coming in and meeting your goals. And there's an evaluative criteria associated with it. That's the difference. Otherwise, it is really just a discussion or a dialogue and an exercise. And so the point of calling it a game is, it makes it fun. And people had an amazingly fun time. There was a lot of laughing, and arguing, and people talking to people that have completely different skill sets from themselves. And so it sort of felt to me like a party. So, there's this aspect of playing, and fun, and having social interaction that's clearly a strong part of it" (anonymous, personal communication, 2014).

The point of calling it a game is that it "makes it fun." While this might be a slip of tongue, it is telling that the project manager specifically started calling it a game because she felt that it provided the context for having fun that other labels didn't offer. Throughout our interview with her, she continued to focus on how best to harness dialogue. The game was not a structure to create play, but in fact, it emerged as a means of capturing and communicating the kind of play that was taking place in these interactions.

According to a serious game designer: "Games have to be fun. No matter how serious, no matter how important or how relevant, if somebody is not having a good time doing it, they're not going to play it. Sounds frivolous, especially if you're talking about serious games, as we are, but it has to be fun" (anonymous, personal communication, 2014). This commitment to fun was common across all the organizations with which we spoke. From designers to the people running the projects in the organizations, the common assumption about games is that they need to be fun and engaging. Fun, for most of the people we spoke to, had a very low bar. It was fun because it was not boring. It was fun because people were attracted to the novelty of the system and were transformed into willing players. The utility of fun is typically associated with keeping people in the system, keeping them engaged. It was never placed in association with a state of play or the freedom that such a state implied. Instead, fun was a means of holding people captive in a voluntary system.

In our conversation with a program director from the World Bank who was using a game to engage youth in community planning, the use of the word *fun* became a limiting factor in the project's life as most of his colleagues only thought of violent video games. "I think social games, social impact games, have gained a little bit more understanding and traction, but still extremely limited. And the word *gamification*, from a business perspective, has kind of opened some space amongst a certain number of colleagues. But a lot still see it as nonserious and nonacademic and kind of a fun extra thing to do as opposed to a way to really engage and motivate young people" (anonymous, personal interview, 2014).

Games applied in civic contexts are difficult to design. But it is far more difficult to design for play. For those looking for playful ways of engaging publics, the language of games and fun dominates the discourse. Through the framework of meaningful inefficiencies, we advocate for prioritizing play over the game that structures it. In the next section, we examine one of our own projects that sought to do just that in the context of a popular social game.

PARTICIPATORY POKÉMON GO

Pokémon Go became a global sensation when it was released in 2016 by the Bay Area game company Niantic. A record-breaking 45 million daily users worldwide were playing the game in the first months after release. As of summer 2018, the game still boasts 5 million daily users, with the numbers remaining steady since 2017. Pokémon Go is an augmented reality game, or a game that uses physical location as the game board where players navigate located data with their mobile phones. In Pokémon Go, players travel to physical locations to capture Pokémon (little fictional creatures) who appear on the map of where the player is standing. Also in the environment are Pokéstops, which are power stations where players collect necessary game resources, located at various "landmarks" in the environment. To play Pokémon Go, players need to move around. In fact, the game rewards movement explicitly by offering 5 kilometer and 10 kilometer Pokémon eggs that hatch only after traveling that distance.

While augmented reality games have been around for over a decade (Gordon & de Souza e Silva, 2011), they have largely been experimental or niche. Even Niantic's first game Ingress, while reaching 5 million active players at its peak, never quite became a household name. Pokémon Go represents the crossover moment where augmented reality entered the mainstream. The reason for this is complicated, but it likely has to do with a perfect storm of simple and engaging game mechanics combined with the rich narrative world of Pokémon that first captured the hearts of all those who grew up with Pokémon games or television, and then quickly spread to those who didn't.

During the peak of its popularity in 2016, we became interested in how the game represented Boston's neighborhoods, and how that corresponded to who was actually playing the game. During the summer of 2016, no matter what neighborhood you went into— rich, poor, Black, White—there were hordes of (mostly) kids wandering the streets with their phones hunting for Pokémon. Players were recognizable through a familiar gait, which included steady oscillation between stopping and running, and the occasional cheer and high five to a stranger catching the same rare Pokémon. The game activated the sociability of city sidewalks—while each was engaged in his or her individual pursuit, there was collectivity and camaraderie that would emerge suddenly and surprisingly. As urbanist William Whyte (2001) said of urban spaces:

> What attracts people most, it would appear, is other people. If I belabor the point, it is because many urban spaces are being designed as though the opposite were true, and that what people liked best were the places they stay away from. People often do talk along such lines; this is why their responses to questionnaires can be so misleading. How many people would say they like to sit in the middle of a crowd? Instead they speak of getting away from it all, and use terms like "escape," "oasis," "retreat." What people do, however, reveals a different priority. (p. 19)

Pokémon Go allows people to escape while being with others. It captures precisely the expressed desires *and* actions that Whyte references. The sociologist Erving Goffman (1959) spoke of "getting away with going away," or the comfort that retreating into a device or a daydream brings when in uncomfortable social situations. While watching young people gather on sidewalks and parks, they would interact, but not be forced to interact. They were in public, but also easily able to go away to their device. This moving in and out of the play space was freedom—an embracing of ambiguity that allowed the player to be both inside and outside of the structures guiding play.

With all the interactions happening alongside the game from a truly diverse group of people, we began to wonder how the data in the game reflected the neighborhoods where players actually lived. Pokéstops are comprised of city landmarks, and recognized landmarks are unequally distributed. While players were coming from all over the city, a truly remarkable feat, the data that comprised the game board in the city of Boston was filled with "Poké deserts," or areas of the city with considerably fewer resources than others. Not surprisingly, these deserts existed in neighborhoods primarily occupied by people of color that lacked other resources as well, including access to healthy foods, adequately maintained parks, and sidewalks.

We reached out to Niantic in the fall of 2016 to partner on a project we called Participatory Pokémon Go that would bring youth voices from these neighborhoods to source and better define Pokéstops in Boston. We launched the project in the summer of 2017 with a series of workshops intended to enable youth to source new Pokéstops and to rewrite the descriptions of existing ones. Early conversations about the nature of the Participatory Pokémon Go project revealed that stakeholders were more interested in focusing on the representation of local history than the objectives of increasing the number of game locations. This new direction was well received by partners in the City and the Boston Public Schools, as well as key allies from grassroots organizations focused on youth development and preservation of local history. Responding to this feedback, we began suggesting an event where participants tour local historic locations and write historically relevant descriptions for Pokéstops.

Initial outreach with community stakeholders involved conversations with offices at Boston's City Hall as well as offices at the Boston Public School headquarters. While the staff at the municipal level had a grasp of salient local issues and needs, it was not until we began engaging with organizations at the grassroots level that we were able to generate unified enthusiasm for the project among all stakeholders. We renamed the project AR (augmented reality) Stories, as interest from youth and partner organizations in Pokémon Go itself was not as high as we originally suspected it would be.

During September we met and coordinated with a group of key stakeholders for the series of AR Stories workshops. We coordinated with 826, a national extracurricular writing program with a branch at one of Boston's premier exam schools in the Dudley Square neighborhood, as a way to engage students who had an interest in writing and research about the topic of local history. We also brought in the Roxbury Historical Society and the Hawthorne Youth and Community Center, one of Roxbury's oldest and most well-respected youth service organizations, to support our work in identifying important historic locations for students to choose from in the creation of their itinerary.

Meetings with key community stakeholders at the grassroots level was not as straightforward as picking up the phone and cold-calling people. Because many local grassroots organizations are wary of collaborating with large corporations and institutions of higher education, given a history of extractive and one-sided collaborations, making connections was a product of reaching out to key gatekeepers to make introductions. Furthermore, conversations required extensive in-person meetings at the headquarters of the different organizations. By conducting frequent face-to-face meetings with stakeholders in the neighborhood where they work, we demonstrated a commitment to being in the community where the project would eventually take place.

We deployed two workshops that engaged students from colleges and high schools in the Boston area. Leading up to the first workshop, we worked with the Roxbury Historical Society to identify a wide range of locations that would benefit from having comprehensive and detailed information included in AR products by Niantic. With this list in hand, we conducted a workshop in early October with Boston students in the 826 Writing program that asked them to review this list and then generate a tour itinerary based on their favorite locations. The itinerary generated by students from this workshop framed our culminating workshop later that month. Leading up to this workshop, we worked with Boston students and the Hawthorne Youth and Community Center to generate a tour guide script based on the itinerary created by students in the previous workshop.

The second workshop started with the tour of historic locations, where attendees of the open house and workshop participants were invited to use Pokémon Go while they took part in the tour. After the tour, workshop participants conducted research and wrote new descriptions for locations they had just visited in the tour itinerary. In addition to the workshop, a table at the open house featured a poster of the locations that were modified or added from the summer program. Attendees were also invited to nominate new locations from the list generated by the Roxbury Historical Society. This event generated six new location descriptions as well as 18 nominations for new locations in Roxbury. In total, the workshops reached over 300 students and recreated 74 Pokéstop locations in Boston.

For this project to achieve local impact, even while boosted by the attraction of Pokémon Go, required investment in local context and an understanding of local barriers to implementation. To speak to the first point, the initial composition of stakeholders for the project consisted of municipal government employees. While the administrators at the city government level as well as administrators of public schools were well positioned to understand the needs of their city, it was not until we started the deployment of the project and interacted with local historical societies and grassroots youth organizations focused on celebrating local culture and history that we began to better understand the nuance of digital equity we had originally set out to address.

Through early pilots of the project and extensive conversations with various stakeholders, we found that digital equity was perceived with urgency through a unique lens, not that of equity in the quantity of locations for game play, but through quality of how locations are represented. Picking up on this interest in how locations are represented, we conducted our own investigation and found that many Pokéstops, including those with historical significance, had little to no writing about them. This clear gap in historical representation revealed itself as a potential turning point in the project that was aligned with the interests of digital equity expressed by stakeholders.

One of the Pokéstop descriptions that was rewritten in the process was "The Faces of Dudley" mural (see Figure 3.2). This mural has, since its creation in 1995, become an important landmark for Black history and empowerment in Boston. However, the original description of the Pokéstop in the game was "BYCC mural, 1995." A group of local high school kids rewrote this description in the culminating workshop and ended up with the following: "This locally famous mural, created in 1995, was remodeled in 2015 to include more women. It has the faces of black activists that made a major contribution to the Roxbury/Dudley area, as it captures the essence of Dudley while emphasizing key black figures to promote pride within the community." This rewritten Pokéstop description is now public in the game.

Figure 3.2. Screenshot of Faces of Dudley mural Pokestop before and after the workshop.

Pokémon Go served as an important and necessary staging area for this expression, but we quickly learned that the project was not about the game—it was about creating conditions for freedom through play. It was about playfulness. The project sought to give young people influence over local data, because otherwise, the freedom of some players would impinge on the freedom of others. Presenting youth with the problem of data and representation was an effective way of making them aware of the field and its limits. We worked through youth-serving organizations to make this happen.

Throughout the project, we encountered tensions between scale and depth. Niantic, of course, was most interested in a city-wide event that captured attention and brought people to the game; the City of Boston and the Boston Public School district were interested in generating goodwill, but as we drilled down into local needs and interest, the real value was in performing local control over the information that increasingly defines urban neighborhoods. It was not simply an invitation to play; it was an invitation to shape the playing field and the sorts of people who would feel welcome to play in the future.

AR Stories was designed as a meaningful inefficiency—one that invited play as a means of creating the conditions for action. It shed light on the complexity of data inequities in Boston, but not just through exposing; it shed light through enabling action. Sicart (2014) argues that play does something more for complex systems: While understanding systems is typically associated with "reduction and synthesis," play brings "action and performance" to systems (p. 97). Play enables action taking, in the Arendtian sense, by allowing people to set things in motion through exploration, experimentation, and discovery. And importantly, this sort of freedom of movement requires trust in the system occupied, trust in the other players, and a willingness to play. The case of AR Stories points to the work involved in designing systems in a civic context that people trust such that they are willing to allow themselves and others to play.

DESIGNING SPACES FOR PLAY

The example of the Participatory Pokémon Go project points to the importance of designing spaces for play, but our discussion of the project thus far has not sufficiently captured the labor involved in accomplishing such a thing. The actual work of designing spaces for play can be classified through the four activities we describe throughout this book: network building, holding space for discussion, distributed ownership, and persistent input. Even if the goal is to produce an app, a service, or an event, the labor involved in making that happen includes building networks, assuring that the right people are at the table, assuring that the project has value once it's complete, and assuring that there are open lines of communication. These are things that practitioners spend their time doing when seeking to design spaces for play.

When we started work on Participatory Pokémon Go, we were eager to build off of existing play spaces, but we did not have strong networks in place. In fact, when the project first started, we deeply undervalued the importance of such networks. Our mistaken assumption was that Pokémon Go was such a popular game that people would be eager to participate simply because of the game's popularity. But that wasn't the case for a couple of reasons: We were particularly seeking participation from youth of color in certain neighborhoods of Boston. While many of the kids in these neighborhoods played Pokémon Go, there wasn't enough of a fan culture among them to respond to the original call to action: "help us make the game better and more reflective of your communities." As Henry Jenkins (2008) has demonstrated, fan cultures can be powerful spaces for activism, but the fan context doesn't manufacture the motivation for action; it merely supports it and gives it structure. The kids were not already motivated to do something about the representation of their neighborhoods, and they certainly didn't see Pokémon Go as a space for activism. As a result, the response to the original call was underwhelming. This response was also the result of where the call to action was coming from. The original competition was housed on a City of Boston website, and it was organized by the authors of this book at the Engagement Lab.

Despite the fact that we have created dozens of similar media interventions in different parts of the world, perhaps because this project took place in our own backyard, we did not place the appropriate attention on assuring the social infrastructure for play. We assumed that because the Pokémon Go player base was so large, players would jump at the chance to play within the new field we were constructing. This turned out to not be the case, most likely because they didn't trust the invitation to play. Playing requires some vulnerability—unless people trust the field, its rules, and the context in which they are playing, they won't enter the game. We needed to do the work of bolstering our network. We reached out beyond the large organizational players, like the City of Boston and the Boston Public Schools, to neighborhood organizations that already have deep ties in the communities we were trying to reach. After introducing the project and having long conversations with community leaders about intentions and desired outcomes, we made the decision to pivot the project so that Pokémon Go was the vehicle, not the destination. People were not enthusiastic about taking their time to make a game better; however, they were enthusiastic about using a game as a tool to make the data representation of their community better. Suddenly we had a growing network of stakeholders who were invested in the project and its outcomes. *The novelty of Pokémon Go was a mechanism to get people's attention, but it was insufficient to get their participation.* The growing network of stakeholders became a well-defined field onto which we could invite people to play.

Bringing people into a space for play *can* be easy. With Pokémon Go's millions of users, people would surely flood into our program. But we didn't want just the average player to participate in our program. We wanted it to be meaningful for youth in particular neighborhoods most impacted by data inequality. It is difficult to distil a mass audience into a specific public. What we have labeled "holding space for discussion" describes the work of getting the *right* people into the designed space. We were seeking to bring in the voices of those most impacted by uneven data representations. Even after cultivating our network (establishing the field), we still needed to create the conditions where people would voluntarily play. It's one thing to work with youth-serving organizations, but if youth participants don't trust the field and its players, then they would not freely play. As Simone de Beauvoir (2011) says, one's individual freedom is contingent on the freedom of others. Exclusion from a play space can actively limit that person's freedom such that the players who are present must operate in "bad faith" as they pretend to be free.

People often assume that with youth media projects, one needs only to get the organization involved and the youth will come along for the ride. And while this can be true in some cases—specifically with school-based programs—presence does not mean participation. While the networks we built helped to get people to the table, we put considerable effort into assuring that the table was set for everyone, and every participant felt comfortable speaking, creating, and dissenting. To do this, we organized a workshop with a small group of high school students to design the session. The time spent recruiting, organizing, and implementing the workshop was considerable, but it produced the results we wanted: a session where youth felt comfortable participating, and full transparency about who was participating and why. Spending the time to get this right turned out to be essential to the project's success.

The two remaining activities, distributing ownership and persistent input, have to do with sustaining play spaces. Throughout the project, we were concerned with assuring that once this particular project ended, participants, both individual and organizational, would be able to continue the work. This was difficult in this particular situation because we were partnering with Niantic, and they had no interest in a long-term engagement. The power to actually impact the game's data was limited to the duration of the project. But, as Pokémon Go was only a vehicle, not the goal, we sought to build the capacity of the organizations we were working with to facilitate similar workshops or activities around augmenting available data about the neighborhood. We spent time talking to people from the administrative side of the Boston Public Schools and other participating organizations to understand what specifically they were interested in. As it turns out, there was little agreement. The youth writing organization 826 was interested in writing for augmented reality, Boston Public Schools was interested in the civic side of data, and the community organizations were interested in ways of activating the youth. We spent hours in meetings with the organizations to understand how the project or pieces of it could be adopted and adapted moving forward.

The work involved in distributing ownership is invaluable, and yet, as we heard from nearly every practitioner with whom we spoke, it goes unrecognized and undervalued. As we discussed earlier, play is social: Even if someone cannot play a particular game at a particular time, as long as a game is understood as playable, and can be theoretically played at a later time, it is legitimized as a play space. Such is true with civic design. Practitioners spend lots of time building the capacity of participants to play; they also spend time assuring that participants can reproduce the play space. As Harry Backlund from City Bureau, a community news organization in Chicago, asked: "Are we going to define our success by the impact that we have directly with the public, or are we going to define it by the changes in the sector that result from the models that we develop (personal communication, 2017)? According to Lissa Soep from Youth Radio, in response to a question about how to assure their direct work with youth has lasting impact, she said, "We want to make it possible for people outside of this building to do some of the things that we do. . . .We need to make it possible for somebody—a group of young people in eastern Kentucky, or a group of young people in St. Louis, or wherever news is breaking and/or wherever there's a story that young people want to tell. We want to be able to provide the tools that allow them to tell that story with the greatest quality and impact" (personal correspondence, 2017). Soep is capturing a common sentiment we heard throughout our interviews: Building the capacity of people to play on their own is the goal of the work and takes up the majority of the practitioner's time.

Finally, persistent input is the process through which players can continue to shape the structure of the field—its boundaries and rules. In the Participatory Pokémon project, we struggled with this because of the limitations of our funding and the reliance on Niantic to actually make the changes within the game. So even as we tried to distribute ownership of the process, we failed to set up mechanisms for the original players to provide input over time. As we learned from our conversations with practitioners, this activity, while widely recognized as essential to civic innovation, is the most difficult to accomplish. We were limited by the cycle of funding— once the project ended, we no longer could pay a project manager to maintain the relationships and continue the conversation about the ongoing work. This is not an unusual problem. However, in retrospect, more of our time during the project should have been spent planning for the project's end. Funders typically do not support such activities, and yet, for play spaces to be trusted, in most cases players need to trust that it will persist beyond the immediate moment.

Participatory Pokémon Go was, like most examples of civic design, imperfectly executed. We made incorrect assumptions and placed resources at times in the wrong places. But, as we have just explained, the project was also successful in many respects. It was not an isolated intervention with discernable outcomes. The process of designing a meaningful inefficiency and its necessary spaces for play was comprised of seemingly prosaic activities directed toward establishing the necessary social infrastructure for the process to exist at all.

CONCLUSION

Meaningful inefficiencies create room for play. But as we have explained in this chapter, the work of designing effective play spaces is complex and difficult. It is not just a matter of designing an attractive field or system; it requires creating the conditions where participants are comfortable and willing to play. This work is not glamorous, and it is typically undervalued or unrecognized. From taking the time to meet with groups to clarify individual and mutual goals, to involving stakeholders in every step of the design process, the actual work involved in civic design is relational and time consuming.

Play is not an aesthetic state or a proximate experience to civic life; it is the fundamental condition in which people can explore, discover, and create in the world—in other words, take action. But no matter how hard one tries, play makes people vulnerable because they submit to rules and allow themselves to be in a state other than the one they occupy normally. As such, play always contains a risk of failure. As Jack Halberstam (2011) writes in his book *The Queer Art of Failure*, "under certain circumstances, failing, losing, forgetting, unmaking, undoing, unbecoming, not knowing may in fact offer more creative, more cooperative, more surprising ways of being in the world" (p. 2). Failure, in the context of play, is twofold: failure of letting oneself play when the rules are not clear (threat of humiliation), and failure within the confines of rules (experimentation and failing safely). The work of building spaces for play is to assure that the former does not come into being and the latter always does. De Beauvoir puts this in the context of ethics. An act is ethical, she says, as long as one creates the conditions for another to pursue her freedom, which often involves the freedom to fail. Playing requires giving oneself over to the possibility of failure, and the sensitivity to others' failures; as such, to play means to be vulnerable. And as civic designers construct spaces for play, they need to do the work of building trust in the systems people inhabit, so when the player fails, there are more opportunities to play. Good games, according to Juul (2013), "promise us a fair chance of redeeming ourselves" (p. 7).

Play is the precondition for action taking. Care is the outcome. In the next chapter, we examine the concept of care and assert that the goal of civic design is to forge conditions in public life whereby people care for shared resources. While engagement specialists pursue outreach and participation, it's not enough. Civic design enables publics to care for the world. In the next chapter, we explain precisely what that means.

[4]

CARE

In April 2018, the MIT Media Lab hosted a hackathon and policy summit focused on improving the breast pump and breastfeeding policy (see Figure 4.1).[1] Entitled *Make the Breast Pump Not Suck*, it brought together hundreds of people who have never participated in, and in many cases, never heard of a hackathon before. Leading up to the event, the year-long planning process sought to distribute ownership beyond the typical stakeholders of innovation and to those rarely consulted, specifically poor women of color. As it was held at the MIT Media Lab, which represents a kind of exclusive access to invention, the organizers spent months building mutual trust so that newcomers would feel comfortable navigating the hackathon. D'Ignazio et al. (2016) suggest that their approach "favors learning over invention, prioritizes listening over ideating, values the production of new social relations over the production of objects." The hackathon was designed with *care*. The organizers were keenly aware of the effort required to care with others in design and to build reciprocal relationships such that participants in systems are not objects, but subjects of their own making.

Make the Breast Pump Not Suck is an example of a meaningful inefficiency. The traditional hackathon is already structured as a game—it has rules, unnecessary obstacles, and a clear goal with competition.[2] The civic hackathon, which has become a mainstay for civic problem solving through coder collectives, city governments, and activist groups, is quite good at building community among coders and practitioners and delivering symbolic

Meaningful Inefficiencies. Eric Gordon and Gabriel Mugar, Oxford University Press (2020).
© Oxford University Press.
DOI: 10.1093/oso/9780190870140.001.0001

Figure 4.1. Breast pump hackathon at the Massachusetts Institute of Technology, April 2018. (Courtesy of Ken Richardson)

value for collaborative problem solving (Schrock, 2018). Most of the time the hackathon is used as an event to corral attention to civic problems—from getting out the vote, to poor conditions of streets, to opioid addiction. Civic hackathons are in the business of getting people to *care about* things. What separated the breast pump event from more traditional hackathons, however, was the attention placed on the quality of play—everything from the design of social spaces to the range of prizes (beyond first, second, and third, awards were named transformation, healthy communities, impact, and superhero).[3] The space was welcoming to nursing mothers in a way that seemed almost contradictory to the techno-modern decor of the Media Lab. The hackathon leaned into the contradiction. On one hand, there was the MIT Media Lab, a hackathon, shiny new tech, and on the other, images of breastfeeding women of color and stories of relationships and political resilience. There was a clear system to play within, but the players were unconventional, and the outcomes were surprising. Beyond inventions, the event inspired an unsuspecting cross section of people to feel comfortable playing in a field to which they were previously not invited.

In the last chapter, we focused on how designers create the conditions for play. In this chapter, we focus on the desired outcome of play: caring. Play enables discovery and commitment to participation in a system. But more than the possibility of individual learning or engagement, play makes it possible for players to care—not just about the system they are in, but about other players, and ideally, the world outside of the system. One outcome of the breast pump hackathon was some clever new technologies, several of which went on to be developed into commercial products. Another outcome, through the deliberate design of lag in the system (gaps in the packed schedule for informal interactions, consistent input from a diverse group of advisors, and considerable effort put into "holding space," or setting the right table for the right people), was caring— *caring for* others, represented by a fundamental acknowledgment of a diversity of experiences, and *caring with* others, represented by participants' collective advocacy for better technology and policy.

Care might seem an abstract goal for a hackathon, especially one with such precise stated outcomes: better tech and better policy. But as with many of the innovations we discuss in this book, focus on outcomes alone does not actually assure sustainable outcomes. Assuring that a group's conclusions acknowledge a plurality of perspectives and realities is essential for innovations in the civic realm to be sustainable. Quick-to-market solutions in civic contexts, such as novel technologies or policies, may generate curiosity, but not care. This chapter makes the case for caring as the answer to the question: Why should public-serving organizations make room for play?

WHY CARE?

While participating in a seminar on civic participation, we were asked by a government employee why civic engagement is valuable for government organizations. Our typical answer had something to do with cost savings. Invest early in process, achieve buy-in from stakeholders, and decisions and processes are more likely to last. But this time, the answer was different. We said something like the following: "Actually, civic engagement is not important, but that people are compelled to engage in something on their own volition is what matters." People who care enough to engage is what designers and civic practitioners should be supporting and cultivating. That rare moment of clarity—engagement doesn't matter, but the intrinsic motivation to engage in something does—brought us down a rabbit hole of philosophical investigations about care, and to a realization that caring is at the core of being in the world, or participating in a social context composed of a plurality of perspectives. This has led us to a logical conclusion that is animating this book: It is the primary responsibility of public-serving organizations to create and support the conditions through which people care for the world.

In the context of smart cities, data journalism, and smart solutions of all sorts, all organizations are compelled to cater to the needs of the user, better and faster. Creating and supporting the conditions through which people care for the world can be aligned with these goals. Local governments creating better and more responsive systems for resident complaints, such as 311, satisfies needs, makes people feel attended to, and can lead to greater trust (Nam & Pardo, 2012). Responsiveness is a baseline element of a public-serving organization's work. But it is only the beginning. Once an organization responds, they must do it consistently and with clarity of motivation. This is where civic innovation happens.

It is common for public-serving organizations to want to motivate constituents or audiences to "care about" something. This phrase describes what people are paying attention to, but it implies no moral responsibility to the thing. One can care about climate change without feeling responsibility to act. According to philosopher Nel Noddings, *caring about* is a state of readiness. "We can . . . 'care about' everyone; that is, we can maintain an internal state of readiness to try to care for whoever crosses our path" (p. 17).

But the actual condition of *caring for* is quite a different thing, she says. It suggests reciprocity, where the one caring and the cared for are mutually engaged in a relationship that leaves each open to discovery. It involves a "stepping out of one's personal frame of reference into an other's" (p. 22). Put another way, Nodding states that when caring for another, "We are in the world of relation, having stepped out of the instrumental world; we have either not yet established goals or we have suspended striving for those already established. We are not attempting to transform the world, but we are allowing ourselves to be transformed" (p. 33). Nodding is building off of the work of the philosopher Martin Buber (1937), who, in his book *I and Thou*, introduces two primary relationships: I–It and I–You. The I–It relationship is the way that one mostly engages in the world, including with other people. I–It positions all things as objects in the perceiver's world, about which the perceiver may care quite a bit or not at all. From the furniture to the technology one uses, to other people on the subway or even close personal friends, one mostly experiences the world as a barrage of relationships with objects to which we give variable attention. I–You, on the other hand, is a fleeting experience of being in relation with another, when the perceiver and the perceived, or the one caring and the cared for, see each other in pure mutuality. Buber describes the I–You as necessarily temporary, a moment of connection that quickly returns to the I–It.

> But this is the exalted melancholy of our fate, that every You in our world must become an It. It does not matter how exclusively present the You was in the direct relation. As soon as the relation has been worked out or has been permeated with a means, the You becomes an object among objects—perhaps the chief, but still one of them, fixed in its size and its limits. (pp. 16–17).

This is a humbling statement from Buber, and it puts into perspective the fickle nature of how people coexist with others. Indeed, even if one is a deeply empathetic person, the experiences of other people are objects in the perceiver's world most of the time. Put simply, relation is necessarily fleeting. But it has to be that way because one wouldn't be able to exist in a constant state of reciprocity. This suggests that "caring work" is difficult—it is a moving target, one not fixable and measurable in the traditional sense.

Caring starts with attention. Drawing users' attention to an issue is necessary. After all, so much contemporary apathy stems from disconnection and cynicism (Cappella & Jamieson, 1997; Gastil & Xenos, 2010; Yamamoto et al., 2017), where people lack belief that any amount of caring about will make a difference. Just because one knows or even cares about an issue does not mean she has the capacity to act. This takes *caring for*, where one's actions emerge in relation to others and are motivated by mutual responsibility between the actor and the acted upon. This isn't a permanent state. Acting never is. What compels people to action is fleeting, much like the action itself. Actions, as defined by Hannah Arendt and as we have explained in previous chapters, are new beginnings, and beginnings always end. The problem then becomes how an organization supports the conditions for action taking. In the last chapter, we introduced play as the precondition for action taking. In this chapter, we explain how organizations support action taking itself. This happens through the manufacture of what Eric Klinenberg (2018) calls "social infrastructure," or the physical and institutional manifestations for social cohesion. Social infrastructure, we argue, is where caring about can transform into caring with. Awareness campaigns can be useful for paying attention, but if that awareness is not given a channel for action, if there are no groups, institutions, spaces, or moments that support action, then it runs the risk of stopping there. This is why organizations are spending time building social infrastructure, creating the conditions through which people can play, so that they might care for each other. Examples like the *Make the Breast Pump Not Suck* hackathon are positive reminders of this. In fact, the lead designer Catherine D'Ignazio admitted that the relationship work, and crafting the social spaces for that to happen, was nearly 75% of her effort on the project (2018, personal correspondence).

There are many caring professions: teachers, social workers, nurses, elder caregivers, first responders, wherein caring relations are primary; where the reciprocity of the one caring and the one cared for is tantamount to success. The professionals in these roles are, under the best of circumstances, operating within a social context that values and supports this work. The civic designers we feature in this book, on the other hand, are typically not operating within an established system of supports because those supports mostly don't exist. Their work, like the caring professional, is to create the conditions for encounter between individuals, but it is also to enable individuals to collectively encounter people and things outside of the systems they inhabit. For example, in the breast pump hackathon, the conditions were ripe for people to care for each other in the space of the event, but more to the point, it extended those caring relations to a collective caring with. Civic design does not end with empathy and relation, because while caring for another is essential for existence (Buber, 1937), it does not itself provide capacity or support to act in the interest of publics (see Chapter 2).

The context for civic design is most often different than the caring professions mentioned earlier. While the teacher cares for the child and in so doing cares for the world, the civic designer cares for the world by creating the conditions for others to care for each other. This is an important distinction for the organizational work we describe in this book. The engagement editor at a newspaper does not typically understand her role as caring for another; instead, she often sees it as transforming an organization so that others may care. If systemic or organizational change is the goal, caring for another directly, or compassionate action, can actually be misleading or even conflicting with larger goals. Hannah Arendt (1968) warns that compassion can sometimes serve as a justification to not act.

> Through compassion the revolutionary-minded humanitarian of the eighteenth century sought to achieve solidarity with the unfortunate and the miserable. . . . Neither compassion nor actual sharing of suffering is enough. We cannot discuss here the mischief that compassion has introduced into modern revolutions by attempts to improve the lot of the unfortunate rather than to establish justice for all. (p. 14)

When an individual act of charity (giving money or volunteering for a day) comes to define the boundaries of political possibilities, it can distance people from the world by making them feel as though they are sympathetically caring for an individual. It preempts action by dividing the world into autonomous actors and dependents. On the contrary, a feminist democratic ethic (Tronto, 2013) moves beyond the standard assumptions of most political theory that assumes the existence of autonomous actors, and instead seeks to explain how individuals balance autonomy and dependency. "Democratic politics should center upon assigning responsibilities for care, and for ensuring that democratic citizens are as capable as possible of participating in this assignment of responsibilities" (p. 30). Tronto calls this caring *with*. This is not simply a value proposition for democracy, or a declaration that vulnerable populations need to be cared for. It is much more profound—all people within a democracy need not only be responsible for caring for others, but also are on the hook for assigning caring responsibilities, for determining what matters, when, and for whom. An act such as voting is a clear assignment of caring responsibility to the person elected, whereas an act such as creating and sharing a political statement on Facebook is not typically accompanied by any expectation of responsibility. A caring democracy, as Tronto puts it, requires a consideration of how each of these actions assigns caring responsibilities, and what kind of social infrastructure is required to make that happen.

Caring, when understood in this way, is distinct from compassion. According to Arendt, when compassion completes the action, when it excludes the cared for in the assessment and assignment of caring responsibilities, then the one caring can turn her back on the world. Caring, says Noddings, "depends not upon rules, or at least not wholly on rules—not upon a prior determination of what is fair or equitable—but upon a constellation of conditions that is viewed through both the eyes of the one-caring and the eyes of the cared-for" (2013, p. 12). And according to Tronto, caring with "consists not of individuals who are the starting point for intellectual reflection, but of humans who are always in relation with others" (2013, p. 36). When Tronto talks of caring with, just as when Arendt talks about worldliness, she does not refer to an isolated individual who contemplates the world; instead, she refers to individual actors who directly engage with a world of relations as a means of being in and of the world.

This might sound abstract, but in practice, it is quite simple. It looks like people prioritizing relations even when there are much faster ways of getting things done. But as we have discovered, this work is filled with complex negotiations. We examine how power and positionality within organizations play into care, and how civic designers are working within this complexity. We return to the last chapter's discussion of play and describe how caring factors into both the experience of being a player and the reflection on having cared. We ask how caring relationships are impacted by the design of play spaces and the dispositions of players. There is a difference between playing the game as designed and playing as an act of resisting the designed system. Each mediates a different relationship to the world and suggests different capacities and limits of caring. The cases we focus on in this chapter are primarily in the space of news and journalism. We look to those who have deliberately, and with some resistance, sought to construct meaningful inefficiencies in the process of news making, with the stated goal of caring with the communities they serve. We pay particular attention to the design of face-to-face encounters to augment the perceived function of the news in a digital era.

POWER PLAY

If caring for is the ability to see an other, to let oneself be subsumed into relation, even if only momentarily, caring with is the ability to see and participate in a world comprised of caring relations. It is the ability to understand, as Simone de Beauvoir says, that one's personal pursuit of freedom impacts others' pursuit of the same, and that the world is not some abstraction to achieve or repair, but a collection of caring relations, each impacting the other. Arendt (1968) talks about this with characteristic clarity:

> For the world is not humane just because it is made by human beings, and it does not become humane just because the human voice sounds in it, but only when it has become the object of discourse. However much we are affected by the things of the world, however deeply they may stir and stimulate us, they become human for us only when we can discuss them with our fellows. (p. 24)

Being in the world is not being in caring relations, Arendt would argue, but in taking actions, where those caring relations are discussed and negotiated with others. This is the antidote for dark times; not just retreating into human connections, but facing them, speaking of them, distributing responsibilities for them, and bringing them to life.

This is hard to reconcile in a contemporary media environment. In the current context of Internet news cycles, it seems contradictory to refer to discourse as anything but distraction. Isn't it the case that contemporary political reality is obfuscated by too much discourse, not too little? When a single tweet from the US President spawns tens of thousands of retweets, hundreds of news stories, and mountains of speculation and interpretation, how can one say that the world lacks discourse? The answer is that this is not discourse, at least not in the Arendtian sense. Retweeting or liking something on Facebook are not in themselves actions. In fact, Siva Vaidhyanathan

(2018) says in his book *Anti-social Media* that they can be just the opposite, that social media is like snack food, addictive and completely lacking substance. Engaging even in political talk on social media does not typically set things in motion; it amplifies. It does not necessarily harness new beginnings; it provides a mechanism for existing messages to persist. Digital content is complex, as its origins are uncanny and often irrelevant, and individuals engage in content as abstracted material, multiple times removed from their human origins. The news cycle is dominated by the purely communicative; in other words, discourse with no referent. Donald Trump's tweets, for example, compel reaction, but primarily on the level of the speech act itself. "How could he have said such a thing?" "Why are those other reporters so upset that he said such a thing?" But the thing itself, whether in reference to mistreated children at the US border or sanctions against Iran, is treated as something to attend to, to care about, but rarely something to care for. In this environment, engaging in the dizzying efficiency of digital discourse is just another way to turn one's back on the world. Breaking through the chatter, seeing the caring relations that comprise the world, is the contemporary challenge.

Civic designers find themselves in a difficult position. How do they take action, or create messages that draw attention to the world, when the landscape of information seems to only want to draw attention to itself? Civic designers are working with a limited resource: not information, which seems limitless, but attention. As Richard Lanham (2006) says, in this new information landscape "attention is the commodity in short supply" (p. xiii). And without attention, there is no care. While all process or communication design is focused on harnessing this limited resource, what distinguishes civic design is the additional need to transform that attention into caring with.

Meaningfully inefficient systems don't simply make people care about things; they give people that occupy systems the conditions to cultivate care and participate in the distribution of caring responsibilities. In our many conversations with civic designers, even if the expressed goal was to design something straightforward—like new housing or a community newspaper—people often spoke of the object of design as the process that empowered people to care. For example, in our discussion with Max Stearns from the City of Boston's Housing Innovation Lab, we were told about the effort to get input into a prototype of alternative housing stock in the city.

> Engagement is about accessing people. And giving people access to the city. But that's the conventional notion of it. I think that what I really learned through [our work] was the value of engagement in the human sense, or when you engage *with* someone. It's like a space to disagree, space to agree, space to ideate and learn from each other, and reflect. (personal correspondence, 2017)

He makes it clear that the traditional approach to communicating about housing innovation was not enough—for the word to actually spread, people needed to know that they cared for something together. And different than seeing a Facebook post with hundreds of likes, the goal was negotiation of values more than attention. Similarly, Mia Birdsong, an activist in Oakland, California, interested in changing the popular narrative around "normal family structure," spoke about her work in this way: "What I would love to be doing is—and it could be with researchers, it could be with journalists, it could be with philanthropy, it could be with policy folks—get a group of you to be thinking about this together" (personal communication, 2017). Getting the message out requires thinking together. This is discourse—when people think together about the world and discuss it with each other. This is markedly distinct from the hypodermic needle model of communication where information is injected into the passive listener (Katz & Lazersfeld, 1955). Civic designers create opportunities for messages to transform through dialogue.

All the people we interviewed for this book could speak about their work through a number of different professional and disciplinary lenses—from outreach coordinator to reporter to data analyst. But we homed in on their unique drive to build relationships, even under adverse conditions. The actual practice of doing this varies quite a bit, which is why we have devoted an entire chapter to describing what this looks like. For civic designers, there is no one path to care. One could argue that *all design* results in the user of an object or process caring about something (Papanek, 2005). For example, for a can opener to be successful, the user needs to understand and care about its function. Same is true with a website or a board game—the designed object hails the user to pay attention to it and to care about it as an object. Caring about is attentiveness to an abstraction. Caring for is relational—one has to see others and be able to imagine them as whole persons outside of any particular situation. Caring with, on the other hand, moves beyond the individual caring relationship and requires a perpetual recognition that the situation exists in context. Consider when Colin Kaepernick, the quarterback for the San Francisco 49ers, took a knee during the national anthem before an August 2016 game. He ignited a national controversy. During an interview after the game, he said: "I am not going to stand up to show pride in a flag for a country that oppresses black people and people of color. To me, this is bigger than football and it would be selfish on my part to look the other way." Kaepernick's actions during this preseason game in August motivated other players to sit down or take a knee during the national anthem for the entirety of the 2017–18 season, ultimately leading to many calls from the President of the United States to fire all players unwilling to stand during the anthem. In August 2018, Donald Trump tweeted the following statement:

> Be happy, be cool! A football game, that fans are paying soooo much money to watch and enjoy, is no place to protest. Most of that money goes to the players anyway. Find another way to protest. Stand proudly for your National Anthem or be Suspended Without Pay!

What's really going on here? To return to our discussion of games in the last chapter, Kaepernick's actions mark a break from the game. They draw attention not to the player, but to the world outside, the injustice of police brutality and systemic racism. While Kaepernick, and those that followed him, were just using a bully pulpit to get out a message, they were also operating in the context of play. This simple act of protest drew attention to the distinction between owners and players, the powerful and (relatively) powerless, and drew attention to the role that players have in questioning the game. That football fans were given permission to care *with* (or resist with) the players as they together looked at racial injustice is the peculiar condition that meaningfully inefficient systems can foster. So, if a game is a metaphor for any system in which people willingly enter in and play, wherein the process, not the outcome, is the reason for play, then the designer of such systems in the civic context needs to accommodate players that step in and out of the system, often as a means of critiquing the system itself.

We spoke with practitioners from government, journalism, civil society, and industry about the design process, and many shared goals of inclusion, relationship building, and participation. But they were coming from institutions with wildly diverse values. The institutional context matters quite a bit (we discuss this in the next chapter). The designed object can never be disentangled from its institutional position. A community organization with strong interpersonal connections will create spaces for play with different goals in mind than a government organization, university, or private corporation. And even within organizations, values are often conflicting. Community groups are motivated by providing services *and* self-preservation, universities are motivated by research outputs *and* impact, government organizations are motivated by representing the public *and* maintaining power, and private corporations, operating with the most clarity, are motivated by profit. So, how does a civic designer, necessarily situated within the constraints of an institutional context, effectively design for care?

The example of Max Stearns from the Housing Innovation Lab in Boston demonstrates a government actor intent on designing a system with inefficiencies to enable people to connect with each other in generative dialogue. And while Stearns was interested in promoting one of the city's agendas of smaller domestic units, he pursued the design process as something emergent, disconnected from necessary or even desirable outcomes. In other words, he embraced the inefficiency despite his institutional affiliation. And while he faced resistance by virtue of representing the City, he also generated trust in the process.

For a completely different organizational context, we spoke to Seed Lynn, a "community storyteller" at the University of Chicago, working on a journalistic project called Southside Stories, which was meant to create opportunities for people in the Southside of Chicago to tell their own stories about their lives. He was keenly aware of his work as embedded within and emerging from institutions. "So, when I say 'we, as institutions,' it's not just 'we, as institutions,' it's 'we, as community.' So, I'm using my institution as a port of entry for community to happen and grow and build" (personal communication, April 2017). He understood his role as representing a large university with contentious relationships in the neighborhood in which he was working, and he sought to embrace the tension as a starting point for dialogue. He eventually left the university because he couldn't reconcile the conflicting value sets between the university (interested in research outputs) and himself as a community organizer (interested in supporting local relationships). From his position in the university, he was acting like a community organizer. This proved to be too difficult, which is why he stepped down from his position and continued doing the work in the neighborhood, but this time as an individual organizer and activist.

But what happens when the designer is not as self-aware as the examples provided here? What happens when the designer uses her institutional grounding as a means of dictating play behaviors, or as Brian Sutton-Smith (1997) puts it, "the landlords usurp the play of their dependents" (p. 101)? This can happen on a small scale, such as when a teacher begins to monitor and regulate the foursquare game on the playground, and the kids then change their play behaviors to comply with a certain set of expectations. They learn to play the landlord's game. Or it can happen on a large scale, as when a "whole culture [is] transformed by such adoptions, as in the effects of African-American jazz and singing in the United States and the world" (p. 101). Sutton-Smith argues that the appropriation of culture requires making room for play in order for the products of play to be captured and used. When the spaces for play are designated by those with power and with considerable interest in maintaining their power, then regardless of how much freedom players have to play, their access to the levers of change outside of the system are limited. So when play spaces are designed by the powerful and intended only to placate, players resist.

In the deliberation game @Stake that we discussed in Chapter 2, we had several examples of people refusing to play the game. The game was used as part of the Participatory Budgeting process in New York City, and it was introduced during the first of six delegate meetings. When the game was introduced in one of the meetings in Queens, two participants, after only a few minutes of gameplay stood up and walked out of the room. At first, we discounted their actions as "spoil sports." But as it turns out, their rejection of the game was an act of resistance to the structures imposed upon them by the perceived landlords. Even though the process was organized by the nonprofit Participatory Budgeting Project, and the game was designed by an academic research lab, it was perceived by those people who walked out to be an uninvited inefficiency imposed upon them by powers that sought to control the process. "We have work to do. Why are we playing a game?" Their rejection of the game wasn't a rejection at all. They were strategically using the game to draw attention to its limitations, and to the assumptions of power and control it imposed upon players.

The rhetoric of resistance is an important caveat to our articulation of caring with. Sutton-Smith argues that play is not one thing or another; it is a set of competing rhetorics, most commonly split between the "rhetoric of progress" (play teaches and builds life skills) and the "rhetoric of fate" (play is the "illusion of mastery over life's circumstances"). These rhetorics are the ways academics, educators, and policymakers speak about the affordances or dangers of play. They are descriptive of the position of the player, as well as those looking at play from the outside. The rhetoric of resistance crosses over these two rhetorics, but it specifically speaks to the experience of the player and the way the player makes sense of the power dynamics of the game. When football players refused to stand for the national anthem in 2017, this was a form of resistant play. They were still playing the game, they were still aware of the rules, but they were simply making obvious the collision between social rules and game rules. As Sutton-Smith (1997) says:

> These two together must generate the non-context-specific higher-order gaming rules, which govern how the players themselves, moment by moment, know how to respond appropriately to what is going on; how to interpret whether a response is nice or nasty; when to be serious; when not to be serious; when to defend friends; make allowances for the young, act deceptively toward those not in one's own group; when the player counts more than winning, or winning counts more than the player. (p. 120)

Not all play is fun. Play is comprised of negotiation, compromise, and resistance. Designing for play is never as easy as crafting a game field. It requires an understanding of all the ways in which players can occupy that field, all the motivations that would bring them there and keep them there, and all the ways that the occasion of voluntary play creates the opportunity for players to care for and with one another.

WHO CARES?

In the summer of 2017, Eric was asked to consult with a state agency on their public engagement strategy. This planning agency functions as a kind of mediator to other organizations who are actually running planning processes. But they play an important role, in that they make decisions about how to distribute state and federal funds to specific municipalities and projects. Their initial concern was that no one knew who they were or understood what they do. What they explained is that they were interested in public engagement as a means of informing the public of the agency's role. When pressed on their reasons for engaging the public, they made it clear that it was difficult for a public agency to make decisions in a vacuum (i.e., no public input), and that they were getting pressure from their boss to "do better public engagement." So two things were going on here: They wanted people to care about their agency, and they wanted to represent to the larger organization that they cared about what the public thought. Sadly, these reasons for public-sector organizations to engage the public in decision making are all too common. When Eric responded with questions about the value of the public caring about the organization, and what leadership really wanted to gain from the appearance of public engagement, they were at a loss. To be clear, these were all well-intentioned people responding to an emerging political context where "engagement" is a buzzword. They were searching for answers, and the consultant was only able to provide more questions. Not surprisingly, that was their last meeting.

Here's the problem. Organizations of all sorts are under increased pressure to "open up" to public input. From news organizations hiring "engagement editors" (Lawrence et al., 2017), to private corporations seeking input on brand identity (Hollobeek, 2011), to governments and universities hiring "chief engagement officers," it is a commonly held assumption that engagement is good. Some of this has to do with changing expectations of "consumers," who, in a world of just in time, same-day access to nearly everything, expect more access to organizations (Gordon, 2017). And some of this has to do with rapidly diminishing trust in organizations; trust in government, media, and nonprofits is falling year after year. In fact, according to the Edelman Trust Barometer, the United States reported the steepest decline in the general trust index ever measured (down 23 points) between 2017 and 2018.[4] Organizations are justifiably worried about this and are responding with a variety of trust-building schemes masquerading as public engagement. But here's what they don't realize: Getting people to care about an organization doesn't mean that they trust it (Nye et al., 1997). Factors such as inequality have a far greater impact on trust than does attentiveness to an issue (Uslaner & Brown, 2003). It would be logical to conclude, therefore, that if trust building was the goal, then the process of engagement would focus on creating a context for equal participation through shared experience, and transparency in process, rather than individual awareness of an organization or issue.

This is easier said than done. Civic organizations, from government to news, are struggling to invent mechanisms to open up to their publics, but quite often they are seeking the easiest, fastest, most novel methods of doing so. The civic designers we spoke to are all engaged in some form of resistance in this regard. While their organizations have charged them with getting people to care about the work, they are often pushing back against accepted procedural norms to create the caring conditions they're after. The actual texture of their work tends to fit within the four activities we have emphasized throughout the book: network building, holding space for discussion, distributing ownership, and assuring consistent input.

Anne Hillman, a journalist who heads up the Solutions Desk of Alaska Public Media, is a powerful example of a civic designer who has effectively employed meaningful inefficiencies to shift an organization's standard practice.[5] Since 2015, she has brought members of the public who wouldn't normally interact into the same physical space, to actively listen to each other, seek understanding, and recognize that conversations won't always lead to agreement or consensus—but remain vital nonetheless. The Community in Unity series began as televised panel discussions with accompanying Q&A sessions, but it later evolved into the community conversation model, ditching the "experts" and focusing on more personal topics of race and identity, and mental health and homelessness.

The project has more recently centered on the topics of incarceration and reintegration of prisoners, by gathering community members, former prisoners, and currently incarcerated people to discuss what life is like in prison and how inmates can fit back into the community once they're released. Hillman has formed partnerships with the Alaska Department of Corrections, along with the inmates and staff of prisons throughout the state.

She sees the project as a tool to reduce the social barriers to reentry and decrease the state's recidivism rate, which is currently 66%. Each curated conversation in the series focuses on a different aspect of the criminal justice system, such as addiction, Alaskan Native culture, and life inside a pretrial facility. The conversations are recorded and broadcasted on public radio and published online.

When asked to reflect on building networks, Hillman made clear that connecting people in the community who think of themselves as separate (specifically members of the public and prisoners) has been the central purpose of the Community in Unity project. Not everyone thought this was true at first. Her partner in the Department of Corrections said that while at first she believed the main goal was educating the public about the state's correctional system, she later realized that the primary goal was actually creating basic human connection.

The Community in Unity series has gathered a diversity of attendees to conversations, ranging from people who have never had experience with the prison system to people who care and are already involved with reentry or criminal justice issues. Hillman managed to forge trusting relationships with superintendents, prisoners, and representatives from the Department of Corrections—the latter so much that she says some people might accuse the department and radio station of crossing an ethical boundary.

Some of the event participants who are involved with criminal justice issues have used the conversations as an opportunity to reach out to people with whom they haven't yet interacted. Hillman has noticed that key event participants—specifically the superintendents of the Fairbanks and Spring Creek prisons—are reaching out and communicating with members of the public without the mediating role of the journalist. She attributes this success to the open nature of the conversations, saying that people are now less nervous to talk to people working inside the correctional system. Volunteers from one of the events have used the initial Community in Unity conversation as a launch pad to communicate with inmates more regularly by planning a three-day event where they can brainstorm next steps and ways to benefit those incarcerated. This is not about building compassion, but about creating a collaborative process, one that begins with empathy, but that results in a distribution of caring responsibilities to the inmates, the community, and the prison staff.

Hillman emphasizes the importance of what we call "holding space," when she explains the effort in creating welcoming and safe social spaces. There are a number of ground rules and structures that Hillman initiates and participants adhere to. Each conversation begins with participants agreeing to actively listen to and respect each other, seek understanding, and accept that issues may arise that can't be fully resolved. Additionally, she recognizes the importance of having good food and drinks for participants to connect over and share, and recognizes the value of sitting in a circle and having community members sit next to inmates. "Just having inmates sit shoulder to shoulder next to community members, I love that . . . It's that part alone right there is very healing to me and very powerful and I think it's healing to the incarcerated people who are joining us in the conversation. It's always interesting because I feel like we always have to tell them before we start, 'Okay, no don't all sit together'" (personal correspondence, 2018).

Hillman will often record the conversations by carrying around the microphone and crouching in front of participants while they speak. This has had the surprising effect of making participants feel more comfortable speaking up because they can maintain eye contact with her and not have to address the whole room at once. Locating the conversations in prisons themselves has been an important choice. In Fairbanks, for example, the conversation took place in the prison gym, where half of the space was being used for housing because of overcrowding.

This attention to the small details of the social space is crucial for Hillman in ensuring that people embrace the openness of the system and never feel intimidated or lost. Hillman and her partners say the project is in part responding to an overemphasis of the victim's voice in normal crime reporting. Instead, they are opening a space where the perpetrators can be heard as well. She says the conversations have primed the institutions to speak with inmates at more regular intervals. Inmates have been given the rare chance to make their voices heard, which is so often stripped away in prison. "What better people to get feedback from than the people who are living in corrections and living in our correctional system," she says. "Their input is the most valuable. They have to live it. I get to go home every day" (personal correspondence, 2018).

According to Hillman, the emphasis on voice and listening is central not only to creating meaningful events but to the broad objectives of rehabilitation and reentry. "When you talk about rehabilitation, they're gonna have all the programming in the world, but sometimes it's just treating people like human beings and like equals. You treat somebody like they're below you or like a monster for their whole life or even a year or six months, how do you expect them to improve and grow as an individual? So I think those human connections, whether it's through basketball or through conversation, I think those are some of the most important parts of rehabilitation" (personal correspondence, 2018).

Hillman has made sure to reach out to key stakeholders before conversations to get their input on what they want from the event. For one event, the team reached out to the superintendent of one of the prisons to help shape the project, by discussing and co-creating the agenda. Before the conversation in Nome, Hillman conducted outreach with people in tribes, people in the city, people from behavioral health, and people from the health corporation to see what they wanted to speak about at the event; she also met with the inmates beforehand to see what they wanted to talk about and what they wanted to learn. "We've definitely built trust with the inmates because we're showing we care and we're listening, but I feel like we also built trust with people who are attending the event and who are listening" (personal correspondence, 2018). Moving forward, there is interest in the Department of Corrections to carry on these conversations in more facilities in the future, even if Hillman cannot participate. This represents the ultimate success of a civic design process. The project has achieved a level of support and ownership from a range of stakeholders that its novelty is transportable to other organizations not as a novelty, but as a normal way of doing things.

Community in Unity is a meaningful inefficiency. It is an attempt to intervene in the way that news gets produced by designing a system wherein a multitude of publics come together with some flexibility to play, to explore, and to encounter the unexpected, as a means not simply of drawing attention to an issue (what the press usually does), but in creating the conditions for people to care. Caring in this case is the ability for people involved to distribute caring responsibilities collectively. It's not just that the people in Nome have built a greater sense of empathy for the inmates, but that through bringing personal experience into shared discourse, they are creating a sense of mutual responsibility for each other. The political philosopher Yascha Mounk (2017) highlights the importance of responsibility for democracy and suggests that its meaning has fundamentally shifted. "In an earlier age, talk of responsibility primarily evoked the individual's duty to help others, it now primarily invokes our responsibility to take care of ourselves ... we have moved from a world in which the conception of 'responsibility-as-duty' predominated to one in which a new conception of 'responsibility-as-accountability' has taken center stage" (p. 30).

Civic designers like Hillman are attempting to reverse this trend, building programs designed to create a sense of duty, not to some abstract sense of service, but a duty to distribute caring responsibilities. Participants in the program, from both sides of the bars, gained value by sharing responsibility for one another and to a shared cause of prison reform. And while the success of the project should be credited to Hillman's drive and persistence, it represents an important institutional framework for assigning caring responsibilities. Hillman's work positioned the public media station in an unsuspecting role: as community builder. She does not talk about her work as content production; instead, she talks about designing spaces for caring that result in good content. All the civic designers we spoke to in the news space were keenly aware of the often frustrating, but productive tension between the institutional priorities of their organizations and their practices of civic design.

This is also true for jesikah maria ross at Capital Public Radio in Sacramento, California. Sacramento is grappling with an affordable housing crisis that is greatly challenging the city's homeless population, middle-class renters, and millennials who might never be able to afford a home. Capital Public Radio, the city's NPR affiliate, has been exploring the history, politics, and economics of housing in Sacramento over the last year with a multiplatform project entitled *The View From Here: Place and Privilege,* which includes a podcast, radio documentary, and online platform for community contributions (see Figure 4.2).

ross, the project lead, has established collaborations with 11 community partners to host events where residents could share their perspectives on the housing crisis in intimate, face-to-face settings. Using the "Story Circle" methodology—which emphasizes the sharing of personal narratives around a common theme—the organizations have engaged wildly diverse members of the public in what ross calls "an experiment in deep listening, radical hospitality, and bridge building" (personal correspondence, 2018).

ross's objectives were twofold: first, to build the capacity of community partners to learn and improve on the Story Circle methodology—culminating in the creation and publishing of a downloadable guide that organizations can use to implement it in future conversations; and second, to host a series of Story Circles that reach more than 100 Sacramento residents from diverse walks of life. The station has supported the work as community

Figure 4.2. Marina Vista public housing residents and Capital Public Radio listeners explore the affordable housing crisis in Sacramento, California, during a Story Circle. (Courtesy of Vanessa Nelson / Capital Public Radio)

outreach and publicity. Station executives, or even those in the newsroom, would likely not state the objectives as such; instead, they talk about the value of outreach and publicity with hard-to-reach communities. ross has been successful at aligning these goals through formal evaluation of her work that actually documents that alignment.

Much of ross's time has been spent on network building. In bringing together 11 organizations working on affordable housing in the region, many of which were aware of each other but hadn't yet collaborated, she had to do the work of aligning stories and goals not only within her own organization but across the partners. For some partners, the Story Circle let them put human faces to the institutions they only knew from a distance. "This is really great because even though I knew them, now I can put a face to a name. I actually have created a contact with that organization, which I think is really helpful to our home base organizations" (personal correspondence, 2018). Together with organizational partners, planning went into attracting a diversity of participation in the Story Circles, in regard to race, ethnicity, occupation, socioeconomic status, and geography. Four conversations took place in four different regions of the city to gather wide participation and diverse perspectives.

The Story Circles have become a gateway for further communication and collaboration among stakeholders, without the mediating role of the radio station. One partner said that after participating in one Story Circle event, she started organizing with other participants around an upcoming ballot initiative on affordable housing. Another participant collected everyone's email addresses at the event so she could include them in an emerging grassroots project. Some participants approached representatives from partner organizations with the intent of joining their ranks. Many of the lasting connections facilitated at the events were due to the programmed networking time the project team scheduled following the conversation.

The project gained traction in large part because of the intentional and strategic structures the facilitators put into place to create a supportive, safe environment where participants feel comfortable sharing personal narratives of struggle and success—in other words, holding space for discussion. According to one project partner, arranging chairs in a circle primes participants to leave their comfort zone and to be patient while interacting with a large group of people.

Facilitators try to make participants feel welcome and comfortable before the conversation begins by introducing themselves to each individual and directing participants to meet each other. ross will often ring a meditation bell and wait patiently for the sound to dissipate before beginning the conversation, creating a sense of unity and anticipation. Facilitators will begin the conversation by sharing their own (pre-rehearsed) story about the topic, as a model for future contributions.

To create a warm environment, the events feature food, candles, and a centerpiece placed in the middle of the circle, such as a bouquet of flowers, to give participants a shared object of contemplation and for them not to get too fixated on facing each other. With large groups, a big conversation will subsequently split off into breakout groups for more intimate interactions. Participants wear ID tags with just their names, not titles, to create an egalitarian mood. The organizers have offered child care at events to make it more feasible with participants with children to attend.

Together these structures create an environment in which participants feel comfortable contributing intimate details and stories about their lives. One project partner said: "Some were sharing their stories and felt so comfortable that they allowed themselves to be vulnerable—they became highly emotional and started tearing up. What stood out to me," according to ross, "was not just that they were sharing intimate details that made them feel highly emotional, but that they felt comfortable doing so, and that the group as a whole seemed to lean in and embrace those who were most vulnerable in that setting" (personal correspondence, 2018).

ross distinguishes the Story Circle model from more traditional discussions, saying that the methodology is designed to surface personal narratives about broad topics rather than rigorous debate about specific political issues, which helps her conceive of both the project's possibilities and its limits. "I think the model, as I understand it, doesn't lend itself to a campaign or to a discussion even of pros and cons. It's not so much a model that's about generating discussion. It's about generating perspectives and a space to hear the perspectives, and make connections, or note gaps, and develop the relationships that can go forward" (personal correspondence, 2018).

The Story Circle model empowers diverse voices to make their perspectives heard. "I forget the man's name, but he spoke about how he was middle management for the state, and very quickly lost everything and became homeless. You mostly hear these stories just walking down the street or in other media places, but to actually have the moment where somebody's sitting in front of you to tell that story was shocking for several people in the circle. Not that we didn't know, not that we haven't heard it, but it was just a moment of reality, that this is a real human being that I'm sitting here talking with and figuring out, 'How did this happen?' We were just all totally embracing that moment of listening to him fully" (personal correspondence, 2018). Creating the conditions for encounter, or caring for, is central to ross's work. But more important, these encounters are designed into a social infrastructure that promotes the assigning of caring responsibilities. The goal is not just to tell stories, or even just to listen to them, but give them the appropriate context where they become actions. The stories function as discourse, as new beginnings, because they are situated in a space designed to accommodate just that.

Additionally, the work of distributing ownership is key. The Story Circle is meant to be a methodology that others can pick up and use in different contexts. They give people the chance to hone their skills in sharing stories and to apply this in different community settings. According to ross, the events are staffed so that people can become "confident in sharing their stories in a public setting," in order to "compel others to action" (personal correspondence, 2018).

As ross describes the multiple layers of objectives associated with the project, she notes that the broad goal is to build trust with and among members of the public and organizations working on affordable housing. The official position of Capital Public Radio would locate the value of this project in reaching out to new audiences. But for ross, that outreach is coupled with a sustainable social infrastructure (physical spaces for sharing stories) that she believes is necessary for local journalism to remain relevant. Like Hillman's story in Alaska, ross civic design process involves building spaces for play with porous boundaries, wherein the game has a perspective, and player behavior is contextualized within it. The work of these designers is to mobilize publics by creating opportunities to play within systems. The way in which those publics play is determined by their relationship to the perspective of the game. Both Hillman and ross understand this and through their design of spaces are doing the difficult work of negotiating the institutional values that house them.

CONCLUSION

This chapter concludes our discussion of the three elements of meaningful inefficiencies—publics, play, and care. Together, they comprise a framework of civic design that is being put into practice across organizational contexts, with startling similarity and consistency. Whether inside a municipal government office or a public radio station, practitioners are pushing up against organizational structures to resist logics of efficiency and to design spaces (physical and virtual) for people to play—discover, encounter, and explore—not for the purpose of fun or attention gathering, but for the purpose of caring. This involves caring for others and then applying those moments of reciprocity to the democratic work of assigning responsibilities for caring. As Hillman and ross demonstrate, news organizations are not limited to simply distributing information; they can also give people what they need to share and distribute responsibilities for one another.

Caring is the product of good design. From can openers to billboards, well-designed objects call out for people to care about them. Civic design, however, is concerned with building capacity for people to not only care about things or even care for others, but to be cared for. It empowers publics to feel responsibility for the equal and just distribution of caring. It's not just the responsibility of teachers, nurses, and social workers, but of everyone to feel responsible that there are enough teachers, nurses, and social workers, and that they are properly supported. The job of the journalist motivated by principles of civic design is not just to distribute information, but to build capacity for people to use, create, and share information.

In the next chapter, we look to the very practical matter of how organizations can support the work of civic design and how practitioners can think about articulating the value of their work. Based on our interviews with practitioners, we developed an instrument called the Reflective Practice Guide (RPG), which is a self-administered reflection tool to help focus in on the specific work involved in civic design. The stories from Hillman and ross in this chapter came directly from the self-administered RPG. We review the tool and discuss how it has been implemented with news organizations and how it might be incorporated into practice more broadly.

[5]

PRACTICE

In 2010, we created an online planning tool called *Community PlanIt*. Much like a game, the platform was structured by three week-long missions and a series of question-based challenges. Each challenge generated points for users, which they could "pledge" to community projects. At the conclusion of the three weeks, the three highest scoring projects won a small cash prize. It was a conversation tool, wherein the user experience was scaffolded within a clear set of rules and overall goals. In fact, it was a game. Over the course of several years, the game was played by dozens of cities for a range of planning projects. People were compelled by the *playful* construct and were eager to have an online component to their planning process. But without fail, someone on the municipal government side would ask how all the conversation data within the platform was being processed to inform decisions. When we explained that conversation, especially conversation produced in the context of play, is not always "good data," they would get confused, or at least disappointed. "The reason we're using this online tool is to aggregate human input to justify government decision making," they would plead. When we explained that the reason to use the online tool was to free people to have *messy*, generative conversations that would lead to greater trust among users so that they might collectively inform good government decisions (Gordon & Baldwin-Philippi, 2014), the disappointment was compounded. *Community PlanIt* was a meaningful inefficiency disguised as an efficient data tool.

Meaningful Inefficiencies. Eric Gordon and Gabriel Mugar, Oxford University Press (2020).
© Oxford University Press.
DOI: 10.1093/oso/9780190870140.001.0001

While it would collect upward of 10,000 comments over a 3-week period, this data was messy, nonsensical (by machine learning standards at the time), and deeply human. The value of the tool, and we didn't fully realize this at the time, was to be a Trojan horse. It sneaked into government board rooms with a false promise of efficiency and then staged a surprise attack with its play-generated discourse to force the consideration of trust building and care as a public-sector responsibility (see Figure 5.1).

This book has provided a framework, both theoretical and practical, for thinking about the unique qualities of a design process that is fundamentally civic in nature. Meaningful inefficiencies seek to build trust and grow capacity of organizational actors to work effectively with publics toward sustainable opportunities to care. The practitioners working in this way, whom we call civic designers, are in large and small organizations, or sometimes individual artists or designers seeking to intervene in larger institutional processes. In whatever organizational capacity they are situated, they create depth in ownership and commitment from users so that

Figure 5.1. Screenshot of Community PlanIt promotional video in Moldova. (Courtesy of Engagement Lab, 2012)

they can push beyond the attention-getting magic of novelty (see Chapter 1). Throughout this book, we have highlighted stories from practitioners in government, news, and community organizations, who are crafting meaningful inefficiencies and, in so doing, pushing back against organizational mandates. In the case of *Community PlanIt*, there was always an internal champion who wanted to use the tool to reshape the value proposition of planning. In some cases, the mandates are entirely unclear, wherein organizations are eager to "do engagement," but absent a clear values statement, they pursue scale and efficiency. These realities put civic designers in a difficult position, as they are not only inventing process as they go, but they also have to justify their work without accepted metrics and the language to speak about outcomes.

For many of the civic designers with whom we spoke, they were alone or on very small teams within large organizations. While some, like those from Youth Radio or City Bureau, are within organizations that have civic design in their DNA, these are still the exception. And even for them, there is pressure to justify to boards or funders that their investment in designing meaningful inefficiencies is worthwhile and productive. In our initial research with practitioners (Gordon & Mugar, 2018), we looked broadly across disciplines and practices, in order to understand the work that people were actually doing. The goal of this research was to establish language that could describe the everyday tasks of the civic designer. That people in government, news organizations, and advocacy groups were thinking similarly about designing with publics demonstrated to us that the principles of civic design were broader than a single industrial or disciplinary standard. However, without structure to connect those practitioners, there is little opportunity for them to develop sustainable practices. Without conferences, journals, meetups, or other mechanism of knowledge sharing, the common practices go unnoticed and uncelebrated (Barab & Duffy, 1998).

The concept of "communities of practice," introduced by Lave and Wenger (1991), explains how learning happens in context and is always situated and collaborative (Vygotsky, 1978). At first, the term was largely applied to educational contexts. In later work, Wenger (1998) connects the term directly to organizational and managerial settings. He explains that communities of practice are formed through three interrelated phenomena: mutual engagement, joint enterprise, and shared repertoire. Mutual engagement is about shared norms formed through a range of discourses and social infrastructure; joint enterprise is a shared understanding of what binds practitioners together (i.e., ethics, meetings, publications); and shared repertoire is a set of communal resources (methods, strategies, instruments, etc.). So while there are practitioners across a range of industries and disciplines that are adopting civic design practices, in order for it to gain legitimacy within any industry or domain, it needs to be localized within mutual engagement, joint enterprise, and shared repertoire.

While previous chapters have focused on mutual engagement (common terms and discourses) and joint enterprise (common ethical framework), this chapter attempts to establish shared repertoire. We focus on how practitioners within specific industries or disciplines legitimize emerging practices of civic design in their work. We begin with a discussion of the creation of an evaluation tool for journalists. We describe how it was used in that specific context and the organizational and institutional challenges it addressed. The evaluation tool makes meaningful inefficiencies actionable, effective, and legitimate. It is, however, just one tool. We do not present it as the solution, but rather as representative of the kind of approach necessary. We share the tool in its entirety and discuss the complexity of its design and implementation. We then discuss a difficult project undertaken by a prison in New England as an example of how a practitioner navigates internal organizational politics to craftily shift programmatic values. Finally, we conclude the chapter and the book by discussing other domains in which civic design might have an impact.

ARTICULATION OF VALUE(S)

Civic designers tend to be driven by a clear set of values, one not always shared with the organization in which they are housed. As such, they tend to have some anxiety regarding their ability to talk to others about the importance of their work. Quantifying engagement, trust, and empowerment is difficult, so the civic designer often has to resort to speaking about apps, clicks, audiences, and other directly observable phenomena (Gordon, 2017). Even as they engage in intricate processes that involve hours and hours of relationship building, they are compelled to report only on the end result, the quality of a story or tool, the number of eyeballs on a website or heads in a room. The actual work they do gets lost because most lack the language to qualify the outcomes and lack the external incentives to qualify the process.

Throughout our research and practice, we have encountered this tension. Practitioners are torn between two masters—publics and employers or funders. As we have described throughout the book, civic designers are supporting and nurturing publics. However, this practice does not always correspond to the bottom line of organizations or the desired return on investment for funders. How does a news organization justify hours of staff time spent talking to people, inviting them to coffee or attending their gatherings, without direct connection to new content? How does a government office support those same activities without directly enhancing service delivery? The problem with civic design is that it tends not to be quantifiable and, as a result, not justifiable. While certainly some of the practitioners we interviewed enjoyed support from their superiors or have had a track record with funders, most have had to creatively interpret (or even misrepresent) their work within their organizations to gain legitimacy.

In 2018 we partnered with the Agora Journalism Center at the University of Oregon to create an instrument (based on our four activities of civic design) for "engagement journalists" to self-evaluate their practice (Lawrence et al., 2019). Engagement journalists (or editors) occupy a niche space within news organizations and are tasked with building relationships with audiences more than they are with generating stories (Lawrence et al., 2017). This group of journalists represents a common situation within the emerging space of civic design in that they sit between two established conventions within an organization—programmatic output (more stories) and public relations (relationship building)—and are seeking legitimacy both inside and outside their home organizations. Our partnership with the Agora Journalism Center was intended to build an instrument to assist these journalists, a civic design community of practice, in articulating the value of the work they do.

We assembled a cohort of seven practicing "engagement journalists" from the United States and Europe. Each of the journalists in the cohort was effectively doing the work of civic design but lacked the language to describe it and the mechanisms to identify the value of the work for news organizations. Over the course of six months, we organized two in-person gatherings and monthly video conferences to discuss each individual's work. Between the first and second gatherings, we introduced an evaluative tool called the Reflective Practice Guide (RPG). This self-guided instrument consists of 14 primary questions to be administered by team members (not an outside evaluator) and an online survey to take after the discussion. Each member of the cohort was invited in because they were working on a project that was specifically oriented toward bringing people together in face-to-face encounters in the context of news and information. Over the course of each project, the practitioners were asked to use the RPG twice as a means of measuring progress.

The members of this initial cohort included seven practicing journalists and academics. They came from the Community Stories Lab in conjunction with Temple University; Alaska Public Media; *The Stand*, a community newspaper in Syracuse, New York; Capital Public Radio in Sacramento, California;

freelance journalists in Lithuania and Germany; and the Bureau of Investigative Journalism in the United Kingdom. All of them were working alone or on small teams, and indeed they were looking for some personal redemption within their organizations or with those organizations with whom they partnered. But beyond personal reward, each was invested in legitimizing or amplifying the work he or she was doing in order to transform the business of local journalism to be interactive, collaborative, and trusted as an integral component of civic life.

The RPG is broken up into four sections organized by the four activities we have been discussing throughout the book: network building, holding space for discussion, distributing ownership, and persistent input. To be clear, these are not merely normative categories; they are descriptive and serve the purpose of organizing the range of practices in which civic designers are engaged. Not everyone at every stage of a project is thinking about persistent input, for example. And if one is at the stage of a project where things just need to get written or built, it may be that network building takes a back seat. The instrument acknowledges that flexibility and uses these categories to contain certain practices, while not mandating their realization at every step. The instrument allows practitioners to document their efforts and then plot them on a chart (see Figure 5.2). The horizontal is a measurement of social infrastructure, spanning from weak on the left side to strong on the right. And the vertical is a measurement of objective, spanning from longevity on the top to novelty below. The goal is to achieve a positive slope, demonstrating progress, because some conditions are more difficult than others, or some contexts are more familiar to the designer than others. Absolute values don't matter, which allows for different sorts of projects, with different levels of embeddedness in a community or mind to sustainability to be valued. For example, novelty can be very useful. It grabs attention and gives people something to talk about, but if a project remains novel over time, it can serve

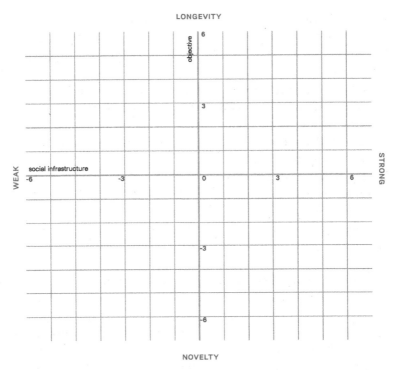

Figure 5.2. Charting progress in Civic Design.

to disrupt publics, as opposed to supporting them. As we discuss in Chapter 1, this is the difference between civic innovation and disruptive innovation. So with the acceptance of relative starting points, the instrument is designed to acknowledge progress, not dictate specific outcomes. As Gordon and Mihailidis state in their volume *Civic Media* (2016), the success of this work is not necessarily in achieving, but in "striving for common good." In other words, it accounts for uneven progress, moments of novelty or experimentation, and strategic shutting down of input to get things built. So, while every project does not need to end up

My New Homeland

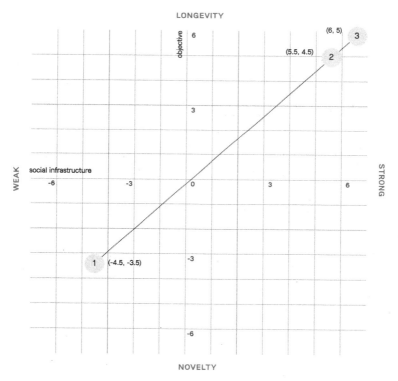

Figure 5.2. Continued.

in the top right quadrant, it *is* imperative that every project results in a positive slope toward that position.

The logic of this progression guided our work with "engagement journalists" and resulted in an instrument that allowed them to measure their progress toward these outcomes, while providing language they could use to articulate the values driving their work and the value of their work to organizations. Following

Community In Unity

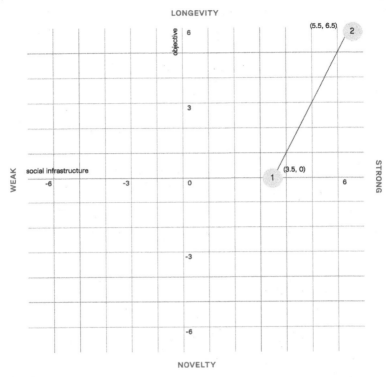

Figure 5.2. Continued.

is the full text of the RPG and a discussion of how it was put into practice.

REFLECTIVE PRACTICE GUIDE

This guide is designed specifically for journalistic media organizations. It should be used to reflect on a specific project. It is important to encourage an open and honest discussion.

Here are some guidelines to follow:

- Gather a group of at least two people involved in a project. This can include anyone working on the project in any capacity, including people at different levels of seniority.
- If you are working alone on a project, try to identify others in your organization who could help you discuss the project. This could be someone who is aware of the project but does not work on it or someone who can simply encourage you to reflect on your own work.
- It would be helpful to assign three roles. First, someone needs to answer the questions. Second, someone should act as a questioner, whose role is to facilitate and encourage the discussion and prompt reflection on interesting insights. Third, someone should act as a note taker, who is less likely to be involved in the discussions as she concentrates on recording responses to the questions.
- This guide is intended to be used multiple times throughout a project. The frequency and timings of this differ from project to project, but we recommend at least two reflection points and up to four. Schedule these with your team according to what is most appropriate for you.

How To Use This Guide

- Gather your reflective team. The reflection activity is split into four main sections; each should take around 15–20 minutes to complete.
- Work through each section together. The person in the question role should read out the instructions and each of the questions in turn.
- Your note taker should record important parts of your answers as you go. These notes will be useful to refer to as you progress through the guide. We'd encourage you to audio record the interview so there is a record of the conversation.

- Every project is different, so not all questions will be relevant to you. It is fine to skip questions. Reflecting on why you skip questions can also be useful.
- Words like "community" and "organization" mean different things to different projects; please see the glossary under "Defining Your Project" to clarify terminology.
- Remember that this is a reflective session, and not a project management meeting, so discussions of successes and failures are equally valid.
- At the conclusion of each reflection, be sure to take the survey (see below).

First: In very succinct terms, define your project and its goals.

Activity #1: Network Building
Since the last reflection point . . .

Describe the new connections you have formed with people, communities, or organizations in support of your project.

- Are any of these connections with community leaders or trusted organizations?
- Are any of these connections with people who have lived experience of the topic of your project?
- How were they formed?
- In what ways do you think they will be useful for the project?

Are there people in your network communicating with each other (without you) who weren't before?

- If yes, how is this happening?
- If you don't know, how can you find out?

Would you feel comfortable reaching out to people in this network in the future?

- If yes, what have you done to achieve this?
- If no, can you imagine things that you can do now to achieve this?

Activity #2: Holding Space for Discussion
Since the last reflection point . . .

What physical places have you created where people can voice their opinions and listen to others?

What digital spaces have you created where people can voice their opinions and listen to others?

- Are the participants in these spaces (physical and digital) broadly representative, or do you feel there are people missing from the discussions, or poorly represented?
- Are there participants from different backgrounds and perspectives? If so, what have you done to support this?
- If you feel that any voices are missing, who are they and what are you doing to address this?

Of the participants in these spaces, do you think that everybody feels able to voice their opinions?

- Have you made any structures or rules (such as ground rules) that make it easier for participants to comfortably contribute?

- How do you allow for constructive disagreement between participants? If you don't, what could you put in place that would allow for this?

How do you demonstrate to the participants in these spaces that you are listening (and in some cases responding) to them?

Activity #3: Distributing Ownership
Since the last reflection point . . .

How have you created new opportunities for people to participate in or shape the project? What are they?

- Can you identify aspects of your project which would benefit from more input?
- How have you created opportunities for a diversity of participants to connect with each other through your project?

What are the elements of your project that can be taken up by people outside of your organization?

- What are you doing to support this?
- Do people have the skills and/or resources needed to effectively participate in the work?
- How can you support the development of skills/resources?

How have you shared the process and outcomes of your work with your project network?

- Has it been effective?

How and where are you sharing the successes and failures of your project with your professional network?

- What other opportunities (if any) do you see to share successes and failures?

Activity #4: Persistent Input
Since the last reflection point . . .

What are your thoughts about what happens when the project ends?

- Beyond continued funding, are there things you are doing or could do to sustain the project's impact?
- What are the challenges faced in planning for the project ending?

Will your presence among the people you've worked with persist for longer than the project duration?

- Have you put anything in place so that they can contact you or your project team after the project has finished?
- Do you feel the project team will continue working with them long term, in some way?

How have you built trust with the people you're working with?

- Do you think this will enable you to continue working to- gether long term?

- Do you feel that this trust will enable you to work with other communities? If so, how?
- Do you feel that your project has created or improved trust among participants and the communities they represent?
- If you feel like you haven't built trust, why do you think this? What could you do differently?

These questions are meant to spark discussion, not simply generate answers. For example, consider the last question: How have you built trust with the people you're working with? There is no way to answer this question succinctly. If approached with the right spirit, the RPG motivates a conversation about whether or not this is a driving value of the work and, if so, what specific things one is doing to realize that value. The RPG is designed to open up as many questions as it resolves, and it is meant to unearth the seemingly banal details that are actually at the core of civic design.

At the conclusion of each reflection, one person from the group is encouraged to take a survey, which then calculates x and y coordinates to plot on the chart.

This survey represents the questions in the RPG and it is meant to be taken immediately after the reflection process. Responses to the survey capture the conversation that happens during the reflection.

SURVEY

1. You and your project team have strengthened your network.
   Not at all 1----------2-----------3----------4----------5 A lot
2. People in the project's network are communicating with each other (without you).
   Not at all 1----------2-----------3----------4----------5 A lot
3. You have created new opportunities for people to participate.
   Not at all 1----------2-----------3----------4----------5 A lot

4. You have made progress in assuring that event participants are broadly representative.
Strongly   Disagree   1----------2-----------3----------4----------5   Strongly Agree

5. Participants feel more comfortable voicing their opinion.
Strongly   Disagree   1----------2-----------3----------4----------5   Strongly Agree

6. There are elements of the project that can be taken up by people outside of your organization.
Strongly   Disagree   1----------2-----------3----------4----------5   Strongly Agree

7. You are more able to share the process and outcomes of your work with project participants.
Strongly   Disagree   1----------2-----------3----------4----------5   Strongly Agree

8. You have shared successes and failures of the project with your wider professional network.
Strongly   Disagree   1----------2-----------3----------4----------5   Strongly Agree

9. You feel confident that the project will continue to have value for participants beyond the life of the project.
Strongly   Disagree   1----------2-----------3----------4----------5   Strongly Agree

10. You are confident that people in the project's network will maintain their connections beyond the life of the project.
Strongly   Disagree   1----------2-----------3----------4----------5   Strongly Agree

11. You and your project team are more able to listen and respond to your participants.
Strongly   Disagree   1----------2-----------3----------4----------5   Strongly Agree

12. You have built more trust with the people with whom you are working.

Strongly   Disagree   1----------2-----------3----------4----------5
Strongly Agree

Coordinates are calculated through a simple algorithm: $X = \sum bQ1 - 5 + 11 - 12$ and $Y = \sum bQ6 - 12$ (see Table 5.1). These x,y coordinates mark a moment in time. That moment is plotted on the chart. When the survey is taken again, the second and perhaps third moments are plotted, and a line is drawn between the points. A positive slope of the line suggests progress toward strong social infrastructure and longevity and serves as a quantified representation of the mostly invisible work of civic design. The recommendation is to display the chart in plain view during the project's duration so as to make one's progress a matter of open discussion.

This graphical representation of progress is not scientific, but dialogic. This is a self-administered survey that is meant to invite conversation among practitioners and between practitioners and managers or funders. In the tradition of empowerment evaluation (Fetterman et al., 2014), which is a participatory approach to evaluation that grew

Table 5.1 CALCULATIONS FOR X, Y
COORDINATES

Survey Response	Variable b
1	−2
2	−1
3	0
4	1
5	2

in popularity in the 1990s and emphasizes capacity building and institutionalization, the RPG provides inroads for practitioners to not simply evaluate past work, but to identify generative actions and explore their sustainability. In the context of emerging communities of practice, where there are often few resources devoted to formal evaluations, this kind of process is not only desirable but necessary.

Over the course of 6 months, as the journalists engaged with the instrument, they were put in the position of having to verbalize the details of their work. They were challenged with seeing the connections between outcomes and process and, most important, they were compelled to confront their minor failures and strategize about how to improve upon them.

The first thing people are asked to do in deploying the RPG is to define the goals of their project. jesikah maria ross from Capital Public Radio, whose story was detailed in the last chapter, explained her "Story Circle" project as follows: "[It] is about improving a tool for conversation, across divides, building trust among people involved in the project, engaging a wider and diverse group of people in conversations about housing and home, and creating this cohort for ongoing peer support" (personal correspondence, 2018). Anne Hillman from Alaska Public Media, also from the last chapter, explained her goal as "helping people connect who have somehow decided they're separate" (personal correspondence, 2018). And other explanations ranged from building trust with audiences to creating stories that matter to people. The participants were asked to define their project and its goals at each reflection point.

The first set of questions has to do with network building. Like all the activities, these questions are meant to unearth details that get lost in the everyday. For example, Sam Ford from the Community Storyteller project in Kentucky, who in collaboration with a local newspaper is focused on sharing everyday stories across the rural county, discovered through his reflection that the majority of their participants were female and that they actually needed methods to bring more men to the table. Other members of the cohort used this opportunity to discuss in detail who they were talking to, what organizations they still needed to talk to, and what kind of interactions they were having. Most people think about networks as core to their strategy. But the second question in the RPG asks whether or not people in the network are communicating without them. This framing reorients the purpose of the network from assisting with the strategic completion of a project to a viable and valuable outcome in itself.

The next set of questions focuses on holding space. These questions are concerned with the design of the space, digital or otherwise, so that people are comfortable and capable of expressing themselves. People tend to use this part of the RPG to reflect on the physical orientation of chairs and bodies if they are thinking about physical spaces, and the connection between social media channels and inclusivity and generativity if they are working online. Ina Daniel from the New Homeland Project in Dortmund, Germany, which puts recent refugees, mostly from Syria, together with World War II refugees to have a conversation, said of her project: "It's important to offer space to utter fears or prejudice and listen to them. In my self-image as a presenter, it is my job to question those opinions, but not to judge them. What are the facts? Is there any personal experience with the issue? Where is the source of news? Are there any ideas for personal constructive solutions?" (personal correspondence, 2018). Through the reflection, Daniel was able to clearly articulate her facilitation process and then use that as a means of assessing her success. Or as ross said about her Story Circle project in Sacramento:

> You need that personal invite. When you're doing mass promotion to people who are already part of the organization, are already listening to Cap Radio, you're just reaching the same people who are self-selecting to be there. And so, the idea is to broaden the conversation and bring people to the table that may not typically be there, so that the project gains more exposure. The model gains more exposure, but also whatever it is that

you're talking about through the model, gains more exposure. (personal correspondence, 2018)

Holding space is a category of activity wherein practitioners can speak about the effort of getting people to the table, and once they're there, creating the opportunities for them to express themselves in the most meaningful way possible. According to Karolis Vysniauskas from Lithuania, holding space is about creating an atmosphere of "coziness and closeness" (personal correspondence, 2018).

Distributing ownership describes the range of activities people engage in to assure that audiences, participants, and contributors feel that they have a stake in the outcomes of any process. For the journalists that used the guide, they spoke about hiring locally, training facilitators that could lead events or forums in the future, and assuring that the benefits of participation extended beyond the event itself. For example, the Southside Photowalk, run by *The Stand* newspaper in Syracuse, New York, made sure that participants left with new skills, new connections, and new networks of people to draw on. In addition to getting people to turn out, they invested time and resources into building capacity and sense of ownership in the process. Maeve McClenaghan from the Bureau of Investigative Journalism in London used the RPG to reflect on a project that used theater to demystify investigative journalism and to connect journalists to refugees. "We met a lot of people that had had experience of the issues we looked at in the play and we made connections, or we deepened connections with the local reporters in each area. And they, in turn, told us that they had made connections with people who had been there on the night and who had since come to them with story tips" (personal correspondence, 2018). McClenaghan spoke about these connections, not as strategic networks, but as people with a stake in the project and with enough trust in the journalists to share their stories. From her initial statement about the purpose of the project, building trust in the institution of journalism was the main goal, and building trusting relationships is the mechanism through which the effort can sustain itself over time.

Finally, persistent input includes all the work put into the longevity of the project, whether it's the way it slips into the banal structures of the organization, or how it puts different communication channels in place between an organization and publics that didn't exist before. This is the most difficult activity for people to wrap their heads around, but with some effort, dialogue around the topic can be very productive. Ultimately, many practitioners do not have the time to think about sustainability beyond the life of a specific project cycle, because they are so focused on getting things done. When they ask the persistent input questions which compel them to think about what happens when the project ends, they can be at a loss. It does seem like a luxury to do long-term planning when one is simply trying to make a thing or create a process. However, what is useful about the reflection is that people begin to realize that they are actually already spending their time doing the work of sustainability. They are doing simple things like setting up Facebook or LinkedIn groups, and they are doing more complicated things like introducing places, activities, and organizations that can sustain activity over time. As jesikah maria ross says, "there has to be an interest in the community to continue to do this and that that's part of the infrastructure that you have to build" (personal correspondence, 2018).

Of course, infrastructure has to be built not only in the public with which one works, but also within organizations. In all cases, if the interest doesn't exist within the organization, then the practice will not persist. The RPG provides a mechanism to reflect on and improve practice, but it also helps to represent that practice to colleagues and superiors in a way that makes sense to them. The axes of social infrastructure and longevity are easily translatable to sustainability. Managers or funders understand that depth of connections and their extension over time will produce stronger programs, more trust, and positively enhance the organization. In some cases, this is precisely why these positions or offices exist. But in most cases, there is a lack of quality metrics to measure success. Certainly, some metrics exist, but they are often misguided or aligned to market values. Metrics might include a 50% increase in Twitter followers or Facebook likes, or increase in meeting attendance or number of local stories. While all of these outcomes may be positive, just because things are measurable does not mean they are meaningful. In fact, Twitter followers can be created through controversy, not trust. Same goes with meeting attendance. And number of local stories does not equal quality, in-depth, or honest reporting. The work of civic design is to enhance the value of civic innovation by redefining how value is measured.

NEW CONTEXTS FOR CIVIC DESIGN

The tool was designed to support designers in navigating these rough waters on their own so that they can effectively communicate within and across organizations. But for it to be effective, it needs to exist within a community of practice. Designers need peers with whom to communicate so that the values work they are doing can be supported and legitimized. The RPG's use within a community of journalists is a best practice. And we can imagine similar groups of practitioners or designers collaborating with this common assessment tool to refine their approach to organizational change. What would this look like in the context of correctional workers? Or health care providers?

In the fall of 2017, we were approached by the leadership of a minimum security women's prison in the Northeast with the request to support a reentry project using virtual reality. The leadership knew that inmates faced anxiety confronting everyday scenarios outside of prison and felt that exposure to virtual scenarios could alleviate some of that anxiety where existing reentry programs were not able to.

Led by our graduate student (at the time), Melissa Teng, we first organized a workshop with prison staff, including nine people from the state and two from community reentry service organizations. We asked them to think about all the pain points that people experience upon release. We sourced over 70 ideas, which we organized and brought to a follow-up workshop with a group of six incarcerated women. The result of this months-long process was three 4-minute 360 videos designed to be viewed in Oculus Go goggles. The videos are meant to anchor conversation with incarcerated women during 1-hour workshops (Teng et al., 2019).

This project might be criticized as mere distraction, or what critical philosopher Herbert Marcuse (1991) would describe as "repressive desublimation." After all, it does nothing to address the underlying structure of mass incarceration and might be seen to enable a broken system by sugarcoating it.

What stands out in this example of civic design in an unlikely context is the work the project leadership engaged in to make the organizations more human and responsive at a time when they are compelled in the opposite direction. The novelty of the virtual reality (VR) pilot brought together a cross section of prison staff and incarcerated individuals to share and collaborate, people who would normally not interact at all, and asked them to "act out" scenarios, play with new ideas, and directly encounter each other. The workshops opened up a space for play, for meaning making, for the reorientation of a public around the shared interest of caring for those reentering society from prison. It also encouraged a conversation among organizational leaders about the value of play and relationships.

In the medical fields, there are existing efforts to resist the hyperrationalization of health care such as the whole health movement (Gaudet & Kligler, 2019). This is an effort within the United States Veterans Administration (VA) that is attempting to reverse the symptoms-based approach to care by asking patients the following questions: "What matters most to you in your life? When you think about these priorities, ask yourself if you are doing everything you can to achieve these goals." This patient-centered care approach is pushing up against the mounds of data and dashboards available about every patient that the physician has at her disposal, and deprioritizing that to elevate the priorities and perceptions of the patient. This is a meaningful inefficiency, and the incorporation of these practices in the institutional mechanisms of health care delivery is civic design. While our research has not brought us to practitioners in the health field, we have no doubt that there are countless civic designers in that field pursuing the kinds of organizational changes we have been describing in this book.

We hope that our attempt at daylighting the practices of civic designers in public-serving organizations will inspire others, from the private sector to the health fields, to validate their own work or seek out those doing this work. As we see it, the allure of efficiency is not going away. Things will get easier and faster, and the entire range of human interactions in the public sphere will be streamlined. But for what? For whom? And why? Civic designers are asking these important questions. But the crushing pace of technological advancement will silence them unless there is coordinated effort to institutionalize their work, provide mechanisms for them to thrive, and ultimately, to build the necessary infrastructure for organizations to cultivate thriving, caring publics in the 21st century.

NOTES

Introduction

1. See National Healthy Sleep Awareness Project (http:sleepeducation.org).

Chapter 1

1. As reported on the City of Boston website—https://www.boston.gov/
departments/new-urban-mechanics/adopt-hydrant

Chapter 2

1. https://www.cnbc.com/2018/04/11/mark-zuckerberg-facebook-is-a-
technology-company-not-media-company.html
2. See https://atstakegame.org/

Chapter 3

1. His best time was 44 minutes from Century City to Los Feliz from 6:00 PM to
6:44 PM.

Chapter 4

1. The program was organized by Catherine D'Ignazio, a research affiliate at the MIT Media Lab and, at the time, a faculty member and principal investigator at the Engagement Lab at Emerson College.
2. https://www.bbc.co.uk/programmes/w3cswgq6
3. https://medium.com/make-the-breast-pump-not-suck-hackathon/who-won-at-the-2018-make-the-breast-pump-not-suck-hackathon-c9f5e3814cae
4. https://www.edelman.com/trust-barometer
5. Hillman is one of seven journalists from the United States and Europe participating in the Finding Common Ground project out of the University of Oregon's Agora Journalism Center.

REFERENCES

Anderson, B. (1983). *Imagined communities: Reflections on the origin and spread of nationalism.* London, UK: Verso.

Arendt, H. (1968). *Men in dark times.* New York, NY: Houghton Mifflin.

Arendt, H. (1998). *The human condition.* Chicago, IL: University of Chicago Press.

Atton, C. (2002). *Alternative media.* London, UK: Sage.

Barab, S. A., & Duffy, T. (1998). From practice fields to communities of practice. Center for Research on Learning and Technology. Technical Report No. 1-98. Indiana University.

Barber, B. (1984). *Strong democracy: Participatory politics for a new age.* Berkeley: University of California Press.

Benhabib, S. (1991). Feminism and postmodernism: An uneasy alliance. *Praxis International, 11*(2), 137–149.

Benhabib, S. (1992). Models of public space: Hannah Arendt, the liberal tradition, and Jurgen Habermas. In Craig Calhoun (Ed.), *Habermas and the public sphere* (pp. 73–98). Cambridge, MA: The MIT Press.

Benkler, Y. (2006). *The wealth of networks: How social production transforms markets and freedom.* New Haven, CT: Yale University Press.

Bjerknes, G., Ehn, P., and Kyng, M. (1987). (Eds.). *Computers and democracy: A Scandinavian challenge.* Brookfield, VT: Gower.

Blumgart, J. (2014, May 26). Shouting to be heard: The case for rethinking public engagement. *Next City.* Retrieved from https://nextcity.org/features/view/the-case-for-rethinking-public-engagement

Bogost, I. (2014). Why gamification is bullshit. In S. Walz & S. Deterding (Eds.), *The gameful world: Approaches, issues, applications* (pp. 65–80). Cambridge, MA: MIT Press.

Bogost, I. (2016). *Play anything: The pleasure of limits, the uses of boredom, and the secret of games.* New York, NY: Basic Books.

Bourdieu, P. (1990). *The logic of practice.* Stanford, CA: Stanford University Press.

Bowker, G. C., Baker, K., Millerand, F., & Ribes, D. (2010). *International handbook of Internet research* (pp. 97–117). Retrieved from http://doi.org/10.1007/978-1-4020-9789-8

Buber, M. (1937). *I and thou.* Edinburgh, Scotland: T&T Clark.

Burstein, R., & Black, A. (2014). *A guide for making innovation offices work.* Innovation Series. New York, NY: IBM Center for the Business of Government.Retrieved from http://www.businessofgovernment.org/sites/default/files/A Guide for Making Innovation Offices Work.pdf

Calhoun, C. (1992). Habermas and the public sphere. In C. Calhoun (Ed.), *Habermas and the public sphere* (pp. 1–50). Cambridge, MA: The MIT Press.

Cappella, J. N., & Jamieson, K. H. (1997). *Spiral of cynicism: The press and the public good.* New York, NY: Oxford University Press.

Csikszentmihalyi, M., & Csikszentmihalyi, I. S. (1992). *Optimal experience: Psychological studies of flow in consciousness.* Cambridge, UK: Cambridge University Press.

de Beauvoir, S. (2011). *The ethics of ambiguity.* New York, NY: Open Road Media.

De Jong, J. (2016). *Dealing with dysfunction: Innovative problem solving in the public sector.* Washington, DC: Brookings Institute.

de Lange, M., & de Waal, M. (2013). Owning the city: New media and citizen engagement in urban design. *First Monday, 18*(11). doi:https://doi.org/10.5210/fm.v18i11.4954

Debord, G. (2000). *Society of the spectacle.* New York, NY: Black and Red.

Deterding, S., Dixon, D., Khaled, R., Nacke, L., Sicart, M., & O'Hara, K. (2011). Gamification: Using game design elements in non-game contexts. In *Proceedings of the 2011 Annual Conference Extended Abstracts on Human Factors in Computing Systems* (pp. 2425–2428). Vancouver, BC: ACM.

Dewey, J. (1922). *Democracy and education.* New York, NY: MacMillan.

Dewey, J. (2012). *The public and its problems: An essay in political inquiry.* Edited by M. L. Rogers. University Park: The Pennsylvania State University Press.

D'Ignazio, C., Hope, A., Metral, M., Brugh, W., Raymond, D., Michelson, B., Achituv, T., & Zuckerman, E. (2016). Towards a feminist hackathon: The "make the breast pump not suck." *Journal of Peer Production,* (8). Retrieved from http://peerproduction.net/issues/issue-8-feminism-and-unhacking-2/peer-reviewed-papers/towards-a-feminist-hackathon-the-make-the-breast-pump-not-suck/

DiSalvo, C. (2012). *Adversarial design.* Cambridge, MA: The MIT Press.

Drew, A. (2017). Living in harmony: The dynamics of social coordination. *Association for Psychological Science.* Retrieved from https://www.psychologicalscience.org/observer/living-in-harmony-the-dynamics-of-social-coordination

Dourish, P., & Bell, G. (2014). *Divining a digital future: Mess and mythology in ubiquitous computing.* Cambridge, MA: MIT Press.

Edelman. (2017). *Trust Barometer—2017. Annual Global Study.* Retrieved from http://www.edelman.com/executive-summary/

Fetterman, D. M., Kaftarian, S. J., & Wandersman, A. (2014). *Empowerment evaluation: Knowledge and tools for self-assessment, evaluation capacity building, and accountability.* New York, NY: Sage.

Fisher, B., & Tronto, J. (1990). Toward a feminist theory of caring. In E. Abel & M. Nelson (Eds.), *Circles of care: Work and identity in women's lives* (pp. 35–62). New York, NY: SUNY Press.

Foucault, M. (1995). *Discipline & punish: The birth of the prison.* New York, NY: Random House.

Fraser, N. (1990). *Rethinking the public sphere: A contribution to the critique of actually existing democracy. Social Text, 25/26,* 56–80. doi:10.2307/466240

Gastil, J., & Xenos, M. (2010). Of attitudes and engagement: Clarifying the reciprocal relationship between civic attitudes and political participation. *Journal of Communication, 60,* 318–343. Retrieved from http://www.la1.psu.edu/cas/jurydem/OfAttitudesAndEngagement.pdf

Gaudet, T., & Kligler, B. (2019). Whole health in the whole system of the Veterans Administration: How will we know we have reached this future state? *The Journal of Alternative and Complementary Medicine, 25*(S1), S7–S11. https://doi.org/10.1089/acm.2018.29061.gau

Gillespie, T. (2015). Platforms intervene. *Social Media and Society, 1*(1). Retrieved from https://journals.sagepub.com/doi/10.1177/2056305115580479

Goffman, E. (1959). *The Presentation of Self in Everyday Life.* New York: Anchor.

Goldsmith, S., & Kleiman, N. (2017). *A new city O/S: The power of open, collaborative, and distributed governance.* Washington, DC: Brookings Institute.

Gordon, E. (2017). *Accelerating public engagement.* Living Cities White Paper. Retrieved from http://engage.livingcities.org. New York.

Gordon, E., & Baldwin-Philippi, J. (2014, February 26). Playful civic learning: Enabling lateral trust and reflection in game-based public participation. *International Journal of Communication.* Retrieved from http://ijoc.org/index.php/ijoc/article/view/2195

Gordon, E., & de Souza e Silva, A. (2011). *Net locality: Why location matters in a networked world.* Malden, MA: Blackwell.

Gordon, E., Haas, J., & Michelson, B. (2017). Civic creativity: Role-playing games in deliberative process. *International Journal of Communication, 11,* 19.

Gordon, E., & Koo, G. (2008). Placeworlds: Using virtual worlds to foster civic engagement. *Space and Culture, 11*(3), 204–221.

Gordon, E., & Lopez, R. (2019). The practice of civic tech: Tensions in the adoption and use of new technologies in community based organizations (CBOs). *Journal of Media and Communication, 7*(3).

Gordon, E., & Mihailidis, P. (2016). Introduction. In E. Gordon & P. Mihailidis (Eds.), *Civic media: Technology, design, practice* (pp. 1–26). Cambridge, MA: MIT Press.

Gordon, E., & Mugar, G. (2018). Civic media practice: Identification and evaluation of media and technology that facilitates democratic process. Retrieved from https://elabhome.blob.core.windows.net/resources/civic_media_practice.pdf

Habermas, J. (1989). *The structural transformation of the public sphere.* Cambridge, UK: Polity Press.

Halberstam, J. (2011). *The queer art of failure.* Raleigh, NC: Duke University Press.

Hamari, J., & Koivisto, J. (2015). Why do people use gamification services? *International Journal of Information Management, 45*(4), 419–431.

Harding, M., Knowles, B., Davies, N., & Rouncefield, M. (2015). HCI, Civic Engagement & Trust. Proceedings of the 33rd Annual ACM Conference on Human Factors in Computing Systems - CHI '15, 2833–2842. https://doi.org/10.1145/2702123.2702255

Hessler, P. (2014, October 13). Tales of the trash: A neighborhood garbage man explores modern Egypt. *The New Yorker.* https://www.newyorker.com/magazine/2014/10/13/tales-trash

Horgan, J. (2016, April 27). Claude Shannon: Tinkerer, prankster, and father of information theory. *IEEE Spectrum Newsletter.* Retrieved from https://spectrum.ieee.org/tech-history/cyberspace/claude-shannon-tinkerer-prankster-and-father-of-information-theory

Huizinga, J. (1955). *Homo Ludens: A study of the play-element in culture.* Boston, MA: Beacon Press.

Israel, B. A., Schulz, A. J., Parker, E. A., & Becker, A. B. (1998). Review of community-based research: Assessing partnership approaches to improve public health. *Annual Review of Public Health, 19*(1), 173–202.

Jacobs, J. (1969). *The death and life of great American cities.* New York, NY: The Modern Library.

Jenkins, H. (2008). *Convergence culture: Where old and new media collide.* New York, NY: NYU Press.

Katz, E., & Lazarsfeld, P. (1955). *Personal influence.* New York, NY: The Free Press.

Klinenberg, E (2013, January 7). Adaptation: How can cities be "climate proofed?" *The New Yorker,* 32–37.

Klinenberg, E. (2018). *Palaces for the people: How social infrastructure can help fight inequality, polarization, and the decline of civic life.* New York, NY: Crown.

Knox, A. B. (2001). Assessing university faculty outreach performance. *College Teaching, 49*(2), 71–74.

Lanham, R. (2006). *The economics of attention: Style and substance in the age of information.* Chicago, IL: University of Chicago Press.

Latour, B. (1998). Where are the missing masses? The sociology of a few mundane artifacts. In W. Bijker & J. Law (Eds.), *Shaping technology/building society: Studies in sociotechnical change* (pp. 225–258). Cambridge, MA.: MIT Press.

Lawrence, R. G., Gordon, E., DeVigal, A., Mellor, C., & Elbaz, J. (2019). Building engagement: Supporting the practice of relational journalism. Agora Journalism Center. Retrieved from http://bit.ly/building-engagement

Lawrence, R. G., Radcliffe, D., & Schmidt, T. R. (2017). Practicing engagement: Participatory journalism in the web 2.0 era. *Journalism Practice, 12*, 1220–1240 https://doi.org/10.1080/17512786.2017.1391712

Le Dantec, C. (2016). *Designing publics.* Cambridge, MA: The MIT Press.

Leighninger, M. (2011). Citizenship and governance in a wild, wired world: How should citizens and public managers use online tools to improve democracy? *National Civic Review.* Retrieved from http://onlinelibrary.wiley.com/doi/10.1002/ncr.20056/abstract

Lerner, J. (2015). *Making democracy fun: How game design can empower citizens and transform politics.* Cambridge, MA: The MIT Press.

Levine, P. (2013). *We are the Ones We Have Been Waiting For: The Promise of Civic Renewal in America.* New York, NY: Oxford University Press.

Levine, R., Locke, C., Searls, D., & Weinberger, D. (2009). *The Cluetrain Manifesto: 10th anniversary edition.* Philadelphia, PA: Basic Books.

Marcuse, H. (1991). *One-dimensional man: Studies in the ideology of advanced industrial societies* (2nd ed.). Boston, MA: Beacon Press.

Marvin, C. (1990). *When old technologies were new.* New York, NY: Oxford University Press.

Matchar, Emily. (2015, April). *Tactical urbanists are improving cities, one rogue fix at a time.* Smithsonian. Retrieved from https://www.smithsonianmag.com/innovation/tactical-urbanists-are-improving-cities-one-rogue-fix-at-a-time-180955049/

McChesney, R. W. (2004). *The problem of the media: U.S. communication politics in the 21st century.* New York, NY: Monthly Review Press.

McDowell C., & Chinchilla, M. (2016). Partnering with communities and institutions. In E. Gordon & P. Mihailidis (Eds.), *Civic media: Technology, design, practice.* Cambridge, MA: MIT Press.

McGonigal, J. (2011). *Reality is broken: Why games make us better and how they can change the World.* New York, NY: Penguin. Retrieved from http://www.amazon.com/Reality-Is-Broken-Better-Change/dp/1594202850

Milan, S. (2016). Liberated technology: Inside emancipatory communication activism. In E. Gordon & P. Mihailidis (Eds.), *Civic media* (pp. 107–124). Cambridge, MA: The MIT Press.

Morison, J. (2010). *Gov 2.0: Towards a User Generated State? Modern Law Review, 73(4), 551–577. Retrieved from 10.1111/j.1468-2230.2010.00808.x*

Mounk, Y. (2017). *The age of responsibility: Luck, choice, and the welfare state.* Cambridge, MA: Harvard University Press.

Muller, M., & Druin, A. (2002). *Participatory design: The third space in HCI.* Cambridge, MA: IBM Research.

Nam, T., & Pardo, T. A. (2012). Transforming city government: A case study of Philly311. *Proceedings of the 6th International Conference on Theory and Practice of Electronic Governance* (pp. 310–319). Albany, NY: ACM.

Nelson, M. J. (2012). Soviet and American precursors to the gamification of work. *Proceeding of the 16th International Academic MindTrek Conference on MindTrek '12*, 23. https://doi.org/10.1145/2393132.2393138

Noddings, N. (2013). *Caring: A relational approach to ethics and moral education.* Berkeley: University of California Press.

Nye, J. S. J., Zelikow, P., & King, D. (1997). *Why people don't trust government.* Cambridge, MA: Harvard University Press.

O'Brien, D. (2018). *The urban commons: How data and technology can rebuild our communities.* Cambridge, MA: Harvard University Press.

O'Donnell, C. (2014). Getting played: Gamification, bullshit, and the rise of algorithmic surveillance. *Surveillance & Society, 12*(3), 349–359. http://www.surveillance-and-society.org | ISSN: 1477-7487

OECD. (2017). *Embracing innovation in government: Global trends 2018.* Paris: OECD.

Oldenburg, R. (1999). *The great good place: Cafes, coffee shops, bookstores, bars, hair salons, and other hangouts at the heart of a community.* Cambridge, UK: De Capo Press.

Papanek, V. (2005). *Design for the real world.* Chicago, IL: Chicago Review Press.

Pew Research Center. (2017). Public trust in government: 1958–2017. http://www.people-press.org/2017/12/14/public-trust-in-government-1958-2017/

Porumbescu, G. (2016). Linking public sector social media and e-government website use to trust in government. *Government Information Quarterly, 33*(2), 291–304.

Rebar. (2011). *The park(ing) day manual.* San Francisco. Retrieved from https://asla.org/uploadedFiles/CMS/Events/Parking_Day_Manual_Consecutive.pdf

Rhinesmith, C. (2016). Community media infrastructure as civic engagement. In E. Gordon & P. Mihailidis (Eds.), *Civic media: Technology, design, practice.* Cambridge, MA: MIT Press.

Robertson, M. (2010). Can't play, won't play. Retrieved from https://kotaku.com/5686393/cant-play-wont-play

Sandmann, L. R., & Weerts, D. J. (2006). Engagement in higher education: Building a federation for action. In *Report of the Proceedings for a Wingspread Conference Establishing the Higher Education Network for Community Engagement (HENCE),* Racine, WI. Retrieved from http://www.henceonline.org.

Schrock, A. R. (2018). *Civic Tech: Making technology work for people.* Los Angeles, CA: Rogue Academic Press.

Schumpeter, J. A. (1939). *Business cycless: A theoretical, historical, and statistical analysis of the capitalist process,* Vol. I. New York and London: McGraw-Hill Book Company.

Shirky, C. (2008). *Here comes everybody: The power of organizing without organizations.* New York, NY: Penguin Press.

Sicart, M. (2014). *Play matters.* Cambridge, MA: MIT Press.

Star, S. L. (2015). The ethnography of infrastructure. In G. Bowker, S. Timmermans, A. E. Clarke, & E. Balka (Eds.), *Boundary objects and beyond: Working with Leigh Star* (pp. 473–488). Cambridge, MA: MIT Press.

Suits, B. (2005). *The grasshopper: Games, life, and utopia.* New York, NY: Broadview Press.

Sutton-Smith, B. (1997). *The ambiguity of play*. Cambridge, MA: Harvard University Press.

Taplin, J. (2017). *Move fast and break things: How Facebook, Google and Amazon cornered culture and undermined democracy*. New York, NY: Little Brown and Company.

Teng, M. Q., Hodge, J., & Gordon, E. (2019). Participatory design of a virtual reality-based reentry training with a women's prison. In *CHI Conference on Human Factors in Computing Systems Extended Abstracts* (CHI'19 Extended Abstracts), May 4–9, 2019, Glasgow, Scotland. New York, NY: ACM. https://doi.org/10.1145/3290607.3299050

The 6 Companies That Own (Almost) All Media. (2017). Retrieved from https://www.webpagefx.com/data/the-6-companies-that-own-almost-all-media/

Thiel, S.-K., Reisinger, M., Röderer, K., & Fröhlich, P. (2016). Playing (with) democracy: A review of gamified participation approaches. *Journal of E-Democracy and Open Government, 8*(2): 32–60.

Tsebelis, G. (1991). *Nested games: Rational choice in comparative politics*. Berkeley: University of California Press.

Tronto, J. C. (2013). *Caring democracy: Markets, equality, and justice*. New York, NY: NYU Press.

Uslaner, E., & Brown, M. (2003). Inequality, trust, and civic engagement. *American Politics Research, 31*. Retrieved from https://www.russellsage.org/sites/all/files/u4/Uslaner and Brown.pdf

Vaidhyanathan, S. (2018). *Anti-social media: How Facebook disconnects us and undermines democracy*. New York, NY: Oxford University Press.

Van der Heijden, H. (2004). User acceptance of hedonic information systems. *MIS Quarterly, 28*(4), 695–704.

Vygotsky, L. S. (1978). *Mind in society: The development of higher psychological processes*. Cambridge, MA: Harvard University Press.

Walz, S., & Deterding, S. (Eds.) (2014). *The gameful world: Approaches, issues, applications*. Cambridge, MA: MIT Press.

Warner, M. (2002). *Publics and counterpublics*. New York, NY: Zone Books.

Weinberger, D. (2015). The Internet that was (and still could be). *The Atlantic*. Retrieved from https://www.theatlantic.com/technology/archive/2015/06/medium-is-the-message-paradise-paved-internet-architecture/396227/

Wenger, E. (1998). *Communities of practice: Learning, meaning and identity*. Cambridge, UK: Cambridge University Press.

Whyte, W. (2001). *Social life of small urban spaces*. New York, NY: Project for Public Spaces.

Yamamoto, M., Kushin, M. J., & Dalisay, F. (2017). Social media and political disengagement among young adults: A moderated mediation model of cynicism, efficacy, and social media use on apathy. *Mass Communication and Society, 20*(2), 149–168. http://doi.org/10.1080/15205436.2016.1224352

Zuckerman, E. (2017). Mistrust, efficacy and the new civics: Understanding the deep roots of the crisis of faith in journalism. Knight Commission Workshop on Trust, Media and American Democracy, Aspen Institute.

Zuckerman, O., & Gal-Oz, A. (2014). Deconstructing gamification: Evaluating the effectiveness of continuous measurement, virtual rewards, and social comparison for promoting physical activity. *Personal and Ubiquitous Computing, 18*(7), 1705–1719. Retrieved from

INDEX

For the benefit of digital users, indexed terms that span two pages (e.g., 52–53) may, on occasion, appear on only one of those pages.

Tables and figures are indicated by *t* and *f* following the page number